Computers
in
Nonassociative
Rings and Algebras

Academic Press Rapid Manuscript Reproduction

Computers
in
Nonassociative
Rings and Algebras

Edited by

Robert E. Beck / Bernard Kolman

Villanova University
Villanova, Pennsylvania

Drexel University
Philadelphia, Pennsylvania

ACADEMIC PRESS, INC.   NEW YORK SAN FRANCISCO LONDON   1977

A Subsidiary of Harcourt Brace Jovanovich, Publishers

ACADEMIC PRESS, INC.
111 Fifth Avenue, New York, New York 10003

United Kingdom Edition published by
ACADEMIC PRESS, INC. (LONDON) LTD.
24/28 Oval Road, London NW1

Library of Congress Cataloging in Publication Data

Main entry under title:

Computers in nonassociative rings and algebras.

"An outgrowth of the special session on computers
in the study of nonassociative rings and algebras . . .
held at the 82nd annual meeting of the American
Mathematical Society in San Antonio, January 22-26,
1976."
Includes index.
1. Nonassociative rings—Data processing—Addresses,
essays, lectures. 2. Nonassociative algebras—Data
processing—Addresses, essays, lectures. I. Beck,
Robert Edward, Date II. Kolman, Bernard, Date
III. American Mathematical Society.
QA252.C65   512'.55   77-5557
ISBN 0-12-083850-8

Contents

List of Contributors

Numbers in parentheses indicate pages on which authors' contributions begin.

Nigel Backhouse (279), Department of Applied Mathematics and Theoretical Physics, University of Liverpool, P.O. Box 147, Liverpool, England L69 3BX

Robert E. Beck (167), Department of Mathematics, Villanova University, Villanova, Pennsylvania 19085

Johan G. F. Belinfante (209), School of Mathematics, Georgia Institute of Technology, Atlanta, Georgia 30332

C. W. Conatser (157), Department of Mathematics, Texas Tech University, Lubbock, Texas 79409

Irvin Roy Hentzel (13), Department of Mathematics, Iowa State University, Ames, Iowa 50010

P. L. Huddleston (157), Division of Science and Mathematics, Edward Waters College, Jacksonville, Florida 32209

Erwin Kleinfeld (1), Department of Mathematics, University of Iowa, Iowa City, Iowa 52242

Bernard Kolman (167), Department of Mathematics, Drexel University, Philadelphia, Pennsylvania 19104

Eugene M. Luks (189), Department of Mathematics, Bucknell University, Lewisburg, Pennsylvania 17837

W. McKay (235), Centre de Recherches Mathématiques, Université de Montréal, Montréal, Québec

J. Patera (235), Centre de Recherches, Mathématiques, Université de Montréal, Montréal, Québec

D. Sankoff (235), Centre de Recherches, Mathématiques, Université de Montréal, Montréal, Québec

B. T. Smith (41), Argonne National Laboratories, 9700 South Cass Avenue, Argonne, Illinois 60439

Ian N. Stewart (167), Department of Mathematics, University of Warwick, Coventry, England

L. T. Wos (41), Argonne National Laboratories, 9700 South Cass Avenue, Argonne, Illinois 60439

Hans Zassenhaus (139), Department of Mathematics, The Ohio State University, Columbus, Ohio 43210

Preface

This volume is an outgrowth of the Special Session on Computers in the Study of Nonassociative Rings and Algebras, which was held at the 82nd annual meeting of the American Mathematical Society in San Antonio, January 22–26, 1976.

The computer has been used in nonassociative rings and algebras for the past 10 to 15 years. However, many of the computational aspects of this work have not been reported. Consequently, researchers involved in similar efforts knew little of what other workers were doing.

Approximately 50 people attended the Special Session and there were fourteen 20-minute papers presented (two by title). The ten papers in this volume are based on papers presented at the meeting.

This volume includes papers that describe algorithmic approaches for solving problems using a computer, papers that describe problems that may be amenable to computer solution, and papers that present data structures and other computational techniques that may be useful in the area of computational algebra. Among the mathematical ideas that these papers discuss are identity processing in nonassociative algebras, representation theory, and structure theory of Lie algebras.

It is the editors' hope that this volume will interest others in the rapidly developing area of computational algebra. They hope that this interest will result in a stronger interaction between the computer scientist and the traditional user of algebraic techniques and will lead to further advances in both disciplines.

We should like to express our thanks to the speakers at the Special Session for their contribution and support of the Session, to those who developed their presentations into papers for this volume for their willingness to work with the editors on this project, to Nancy Cressman for typing the papers in final form, to Anil Agarwal for the necessary art work, and to the staff of Academic Press for their interest and cooperation.

EXAMPLES, COUNTEREXAMPLES AND THE COMPUTER

Erwin Kleinfeld

The following is a personal account of one mathematician's experiences in attempting to use computers in his research. There are any number of factors which might discourage a person from making the attempt, as well as negative experiences when one finally takes the plunge. Perhaps others can benefit from this experience and come to a fruitful decision on their own problems. Actually the opportunity and desire to use a computer has occurred on just two occasions. The first happened under almost ideal circumstances. While awaiting clearance, which never came through, I was given the assignment by the project director to devise some mathematical problem that could be solved on a computer (SWAC). This removed the usual problem of cost and availability of programmers, but then I was completely ignorant of what could be accomplished in this way. The problem I finally settled on was to determine whether there exist non-Desarguesian projective planes of order 16 (with 17 points on each line). Incidentally, the question is still open for order 10, despite large numbers of computations by experts, but the problem of order 16 is a much easier one, since 16 is a primepower while 10 is not. The reader who wishes to find out how it was done need only consult [6] for the details. The idea was to calculate Veblen-Wedderburn systems of a certain type, which SWAC dutifully carried out. If one tries to do that in the crudest possible way, one finds the computations too combersome and so, it was necessary to devise a clever scheme to reduce the problem to manageable proportions. The most exciting

1

moment came when the computer printed out all the examples, for it indicated that the problem was solved and that there had to be at least one non-Desarguesian plane. But months later when the euphoria had worn off, the question of what to do with all the IBM cards presented itself and now it was clear that the work had just begun, because there were obvious questions the computers hadn't settled. This would now have to be dealt with before the information could be published, for it was necessary to determine which systems were isomorphic and how many distinct planes they came from. By then it was no longer easy to have access to a computer and programmers and so the work was put aside until the desire to publish a paper became more urgent, which as it turned out, resulted in a time lapse of several years. Actually, nice things happened as a result of having completed the work, for one of the examples the computer discovered turned out to be more interesting than the others and led to an infinite class of planes that were later exploited in [3]. Also, Knuth was able to settle the question of existence on non-Desarguesian planes of the remaining orders of order a power of two in [7].

The second time an opportunity to use a computer arose concerns an identity in alternative rings. A ring may be defined to be alternative in case every subring which can be generated by two elements is associative. The associator is defined by $(x,y,z) = (xy)z - x(yz)$, and the commutator by $(x,y) = xy - yx$. It had been discovered that in every alternative ring the identity

$$(1) \qquad ((w,x)^4,y,z) = 0,$$

holds, but the burning question remaining at that time was whether in fact $((w,x)^2,y,z)$ might be zero. The reason why one might hope for such an identity to hold is that in fact it has been shown that

$$(2) \qquad (w,x)((w,x)^2,y,z) = 0 = ((w,x)^2,y,z)(w,x),$$

so that at least under the absence of proper divisors of zero this

might be true. In addition, Dorofeev had shown [1] that
$((w,x)^2,y,z) = 0$ in every alternative ring that could be generated
by three elements, while showing [2] that a related but decidedly
stronger identity did not hold in alternative rings with six
generators. There are several reasons for wanting to know the
answer to this question and this has been discussed elsewhere
[8,4]. Eventually a counterexample to the identity was produced
by the author and his wife [4]. The actual example takes half a
page to write down. The casual reader would be unlikely to guess
at the trials and tribulations that went into producing this
example. We shall remedy this by producing the background of this
example, show how it was arrived at and incidentally produce a
more general example that was hinted at in [4] which is valid for
all characteristics. Incidentally, mathematicians seem to be dis-
tinctly prejudiced against counterexamples. How often do we allow
the publication of important examples and counterexamples?   In
failing to prove desirable theorems how often do we place a higher
value on a result with "suitable hypothesis" added that will make
the result come true? I admit to those same prejudices at various
times in the past, although I tend to regret them now. Could it
be that we are afraid of some brilliant child or amateur of our
imagination who will prove capable of producing astonishing
examples and could thus threaten our status? If so, my personal
experiences have shown otherwise. You have to be just as well-
informed and diligent to produce the example or counterexample.
Once you have found it though, you are not supposed to dwell on
how you found it, just on why it works and what consequences it
has. At least in this paper I shall remedy this situation by
giving a detailed account of how this example was obtained. Since
I have admitted to being a victim of these prejudices in the past,
I will also admit to having made a number of unsuccessful attempts
to prove the non-identity in question before it ever occurred to
me that one should attempt to find a counterexample. When we did
find the counterexample, we reduced it to its barest essentials

so that it appears as though we had pulled the rabbit out of a hat,
which was not the case. There is a fool-proof finite method of
deciding whether any identity of this type holds in an alternative
ring and it will be fruitful to explain the procedure, as it will
aid in following the various paths we took. One simply decides
to study the free alternative ring on a finite number of generators. To study the non-identity in question, one can make do with
four generators: w, x, y, z. The free alternative ring can be
thought of as consisting of sums of words of certain lengths,
obtained from the free non-associative ring having the same generators. To convert the free non-associative ring into the free
alternative ring, one need only impose the relations

(3) $(a,b,c)=(b,c,a)=(c,a,b)= -(a,c,b)= -(c,b,a)= -(b,a,c),$

for all words a,b,c. Note well that (3) sets up a linear dependence relation among words of equal length, no matter what words
are substituted for a,b and c. So we can think of the free alternative ring as having linear dependence relations among the words
of the free non-associative ring and now the object of the game
is to determine a basis for words of equal length, so that we can
determine what the identities are. One quickly learns also to
set to zero those words which do not affect the outcome of the
identity. Thus, in studying $((w,x)^2,y,z)$ we can set to zero all
words that contain more than one y or more than one z or more than
two w or more than two x. Clearly, then we need a basis for words
of length six, subject to these economies. Even with the new conditions this will tell us whether the eight words of length six in
$((w,x)^2,y,z)$ are related by the identity or not, when we express
each word in terms of the basic elements. To obtain a feeling for
the problem one should write out first all the words of length two
in the free ring on four generators.

L2:

w^2	wx	wy	wz	xw
x^2	xy	xz	yw	yx
y^2	yz	zw	zx	zy
z^2				

just to verify that there are sixteen. There are no relations between them in the free alternative ring. However, one can make a mental note to put y^2 and z^2 equal to zero, for reasons that we have already discussed, as they will not affect the outcome. When we come to length three, we observe that there are 128 words to start with. However, those involving only two letters or one need not be written in two different associations, since the alternative law states these will be equal. Then we can eliminate all words that do not affect the outcome. It can then be worked out that from the remaining words the following can be selected as a basis.

L3:

$(xy)z$	$(yz)x$	$(zx)y$	$(xz)y$	$(zy)x$
$(yx)z$	$x(yz)$	$(wx)y$	$(xy)w$	$(yw)x$
$(xw)y$	$(wy)x$	$(yx)w$	$w(xy)$	$(wx)z$
$(xz)w$	$(zw)x$	$(xw)z$	$(wz)x$	$(zx)w$
$w(xz)$	$(wy)z$	$(yz)w,$	$(zw)y$	$(yw)z$
$(wz)y$	$(zy)w$	$w(yz)$	w^2x	wxw
xw^2	w^2y	wyw	yw^2	w^2z
wzw	zw^2	x^2w	xwx	wx^2
x^2y	xyx	yx^2	x^2z	xzx
zx^2				

We observe that there are 46 basis elements for words of length three. The linear dependence relations can be obtained readily from the use of (3), but the details will not be spelled out here. At this stage the computations have not yet gotten out of hand, but at the very next step, in attempting to handle words of length four, the project ceases to be fun and one begins to wonder if this

might not be ideal work for a computer. We did consult some
experts at this point who were quite discouraging on the ability
of a computer to carry out this work to its logical conclusion and
so we were forced to rethink the problem. Surely if we were to set
more words to zero at an earlier stage it would be easier to carry
on the project, but then if it failed to produce a counterexample
the effort would have been inconclusive and so really a waste of
time. We decided to go ahead with this to see how much work this
would entail. The rationale behind the choice of words that would
be set to zero stemmed from the thought that it might be sufficient if only two of the words of length six were to survive,
while deliberately setting out to make the others zero. This could
be accomplished by reducing the original list of words of length
two to the number seven, namely

$$L'2:$$

w^2	wx	xw	wy	yw
wz	zw			

This of course destroys the symmetry between w and x, but there
was no reason to suppose that need be bad. Interestingly enough,
the basic elements for words of length three were now reduced to
the following 19.

$$L'3:$$

w^2x	wxw	xw^2	w^2y	wyw
yw^2	w^2z	wzw	zw^2	xwx
(wx)y	(yw)x	(xw)y	(wx)z	(zw)x
(xw)z	(wz)y	(yw)z	(zw)y	

We omit the dependence relations for all the other words of length
three. We carried out the work for length four and found a basis
of 29 elements and then a basis of 22 elements for words of length
five. These figures don't sound bad at all, except that the reader
must remember that one had a total of 113 non-zero words of length
four to catalogue, and 161 non-zero words of length five, together

with all the dependence relations. The bookkeeping at this point became so difficult, plus the consideration of having made errors or about to make errors such a strong possibility that we again felt in need of rethinking the problem. I wish that I could say that we persevered and carried out the work one step further to length six and produced the counterexample, but that is not what happened. We consulted again with some experts about the possibility of using a computer to finish the work, but again they discouraged us and so the problem rested in limbo for another while. Perhaps in time we would have resumed the effort, but fate intervened to show us a much more satisfactory way to produce the counterexample. To this day we do not know whether our previous method would have led to a counterexample and of course it would not be very productive to find this out now. What led us to this counterexample could only be described as serendipity. We had been working on a second problem dealing with possible extensions of the Cayley numbers [5]. In this context we found that if indeed $((w,x)^2, y, z)$ were zero, that this would help us prove the result we were looking for. The clue to the resolution of this dilemma came from the way the substitution was going to be made into $((w,x)^2, y, z) = 0$, for it turned out that w was going to be an idempotent. That was all the clue one needs to reduce the counterexample to manageable proportions, for one has the Peirce decomposition in an alternative ring to work with. That is to say $z = a_{11} + a_{10} + a_{01} + a_{00}$, where the subscripts tell us whether the idempotent e is going to be a right or left annihilator or identity, just as in the associative case. So we made the decision that we would work with five generators (an added measure of insurance) and that w would be in A_{01}, with x, y and z in A_{10}, with the idea of studying $((e, w-x)^2, y, z)$. There is of course an extensive literature on the behavior of such elements in an alternative ring that one can draw on. In particular, it is known that $a_{10}^2 = 0 = a_{01}^2$, and that the choice of w − x really represents an arbitrary element of A. Calculating $(e, w-x)^2 = (-x-w)^2 =$

wx + xw, an element of $A_{00} + A_{11}$, this dictates that y and z need both be in either A_{01} or both be in A_{10}. It is perfectly arbitrary to put them both in A_{10} as we did. At this point, one can go ahead and form the free ring on generators e,w,x,y,z, subject to w in A_{01} and x,y,z in A_{10}. The happy circumstance is that the decisive computations now involve words of length at most four, since we are trying to determine the fate of $((e,w-x)^2,y,z) = (wx+xw,y,z)$, rather than length six, a considerable improvement in reducing the problem to manageable proportions. As it turns out, we obtain the following bases.

L1:

e	w	x	y	z

L2:

wx	wy	wz	xw	xy
xz	yw	yz	zw	

L3:

(xy)z	z(xy)	(xy)w	(yw)x	(xw)y
(yz)w	(zw)y	(yw)z	(xz)w	(xw)z
(zw)x				

L4:

[(xy)w]z	[(xz)w]y	[(xy)z]w	x[(yz)w]

Of course all words in which any letter is repeated can be set to zero because it does not affect the outcome of the words in question. Because it is awkward to write the words of length four, we have shortened them to the numbers 1,2,3,4, where 1 = [(xy)w]z, 2 = -x[(yz)], 3 = [(xz)w]y, and 4 = [(xy)z]w. The multiplication table is now completely determined. In the accompanying table we produce those results. Whenever a product is zero, we have left the appropriate space blank, to save writing. The result is an algebra of dimension 29 in which (wx+xw,y,z) = 1 + 2 -

3 + 4 + 4. In the original example [4] we used the further
relations (xy)w = (yz)w = (xz)w = 0, which reduced the dimension
from 29 to 23 and eliminated 1,2,3, so that characteristic different from two was essential, since the critical identity now
equalled 4 + 4. The lesson we learned from this is to avoid the
computer when suitable alternatives exist. I might also add that
it was lucky we were working on the second problem, which gave us
the important clue. Eventually, if this hadn't worked, we would
have come back to the original method and been forced into much
nastier computations. The irony is we did finally avail ourselves
of the computer in checking that the 23 dimensional algebra really
is alternative. That of course is easy to program and takes but a
short time to compute, but is too tedious if one has to do it one-
self. Besides we were mentally exhausted at this point and eager
to have an independent check on our work. There is one consequence
of this non-identity worth mentioning. It implies the existence
of a nonzero nilpotent ideal in the free alternative ring on four
or more generators [4].   Even though we spent a great deal of
time and effort to produce a counterexample which can be explained
in a single page and even though part of the effort proved unnecessary, we learned a great deal in the process and felt quite
pleased to have settled this problem.

References

[1] G.V. Dorofeev, *Alternative rings with three generators,*
     Sibirsk. Mat. E. 4(1963), 1029-1048.

[2] G.V. Dorofeev, *An example in the theory of alternative rings,*
     Sibirsk. Mat. E. 4(1963), 1049-1052.

[3] D.R. Hughes and Erwin Kleinfeld, *Semi-nuclear extensions of
     Galois fields,* Amer. J. Math. 82(1960), 389-392.

[4] Margaret M. Humm and Erwin Kleinfeld, *On free alternative
     rings,* J. of Comb. Th. 2(1967), 140-144.

[5] Margaret M. Humm and Erwin Kleinfeld, *On extensions of
     Cayley algebras,* Proc. Amer. Math. Soc. 17(1966), 1203-1204.

	e	w	x	y	z	xy·z	z·xy	xy·w	yw·x	xw·y	yz·w	zw·y	yw·z	xz·w	xw·z
e	e		x	y	z		z·xy	xy·w	yw·x	xw·y	yz·w	zw·y	yw·z	xz·w	xw·z
w		w	wx	wy	wz	-1-2+3-4									
x		xw		xy	xz								-2	1+4	3-4
y		yw	-xy		yz									-3	2+4
z		zw	-xz	-yz			-1		-3+4	-2-4					
xy·z		4													
z·xy	z·xy														
xy·w					1										
yw·x					3-4										
xw·y					2+4										
yz·w			2												
zw·y			-1-4												
yw·z			-3+4												
xz·w				3											
xw·z				-2-4											
zw·x				1+4											
wx															
wy															
wz															
xw		xw		xw·y	xw·z										
xy		xy	xy·w		xy·z										
xz		xz	xz·w	-xy·z											
yw		yw	yw·x		yw·z										
yz		yz	yz·w	xy·z											
zw		zw	zw·x	zw·y											
1		1													
2		2													
3		3													
4		4													

zw·x	wx	wy	wz	λw	xy	xz	yw	yz	zw	1	2	3	4
zw·x				xw			yw		zw				
					−xy·w	−xz·w		−yz·w					
		xy·w +xw·y	xw·z +xz·w					z·xy					
−1−4	yw·x −xy·w		yz·w +yw·z			−z·xy							
	−xz·w +zw·x	zw·y −yz·w			z·xy								
								−1 + 3−4					
					1+2+4								
				−2+ 3−4									
									1+4				
						3−4							
			2+4										

[6]   Erwin Kleinfeld, *Techniques for enumerating Veblen-Wedder-burn systems*, J. Assoc. Comp. Mach. $\underline{7}$(1960), 330-337.

[7]   Donald E. Knuth, *Finite semifields and projective planes*, J. of Algebra, $\underline{2}$(1965), 182-217.

[8]   L.A. Skornyakov, *Alternative rings*, Reconditi di Mat. $\underline{24}$ (1965), 1-13.

Department of Mathematics
University of Iowa
Iowa City, Iowa   52242

PROCESSING IDENTITIES BY GROUP REPRESENTATION

Irvin Roy Hentzel

1. Introduction

We present here a technique which enables a computer to
process identities. The examples used are from nonassociative
algebras, but the method is general and can be applied to many
other situations. Identities are expressed as elements of the
group ring on the symmetric group and then are transferred to
matrices by the well-known isomorphisms of group rings. This
approach has several advantages over the standard approach, which
uses a system of linear equations. First, it indicates whether
a desirable identity is a consequence of other assumed identities.
If the identity is not a consequence of the assumed identities, it
tells how close it is to being true. This lets the investigator
know which routes are worth pursuing. Second, it uses very little
array area. Identities involving four unknowns are processed with
the largest array being a 3 × 3 matrix. Five unknowns require a
6 × 6 matrix, and 6 unknowns require a 16 × 16. A 16 × 16 matrix
seems large; remember, however, that a system involving 720 unknowns
is being solved with an array having 1/3 that many entries. The
traditional approach would be to use a 720×720 matrix. Third, our
approach can be modified easily to suit various situations. Fourth,
our approach breaks identities down to their minimal parts; this makes
ranking identities easier. It also makes it easy to see exactly
what is required to prove a result. Without a technique such as
this, identities of degree four are about all that can be managed
effectively. The group ring approach offers an effective way to

13

process identities of larger degree.

2.   Theory

We shall base our work on nonassociative algebras over a
field F. We do not require that our algebras be finite dimensional. We will require in the applications that the characteristic of F is not 2 or 3.

Let R be a nonassociative algebra over a field F.   Let

$$f: \prod_{i=1}^{n} R \longrightarrow R.$$

An identity of R is a sum

$$\sum_{\pi \varepsilon S_n} \gamma_\pi f_\pi$$

which takes the value $0 \varepsilon R$ for all choices of its arguments.
S_n is the symmetric group on n objects, γ_π is a scalar in F, and
by f_π we mean to permute the arguments by π before f is evaluated
on them. We call the element

$$\Phi = \sum_{\pi \varepsilon S_n} \gamma_\pi \pi$$

of the group ring of S_n over F, an identity of R based on the
function f if $f_\Phi = \sum_{\pi \varepsilon S_n} \gamma_\pi f_\pi$ is an identity of R. The set of
all identities of R based on the function f form a left ideal of
the group ring.

If $\Phi_1, \Phi_2, \cdots, \Phi_k$ are all identities of R based on the function
f, we are interested in the identities that these imply. We say
an identity Φ is implied by $\Phi_1, \Phi_2, \cdots, \Phi_k$ if by varying the arguments in $f_{\Phi_1}, f_{\Phi_2}, \cdots, f_{\Phi_k}$ and adding them together, one gets f_Φ.
This process, however complicated it may appear, is easily des-

cribed and dealt with. Φ is a consequence of $\Phi_1, \Phi_2, \ldots, \Phi_k$ if and only if Φ is in the left ideal generated by $\Phi_1, \Phi_2, \ldots, \Phi_k$.

Given two sets of identities of R based on the function F, they may be compared by examining the left ideal they generate. In particular, they are equivalent if and only if they generate the same left ideal.

An identity is called minimal if the left ideal it generates is a minimal left ideal of the group ring. If the characteristic of F does not divide n!, then the group ring is semisimple, and every left ideal is the direct sum of simple left ideals. This means that every identity of R based on f is equivalent to assuming one or more minimal identities of R based on f.

3.   Identities Involving Several Functions

The previous technique was based on the assumption that an identity could be expressed by a single function. Many common identities, like the ubiquitous Teichmuller identity which holds in any nonassociative algebra,

(1) $(ab,c,d)-(a,bc,d)+(a,b,cd)-a(b,c,d)-(a,b,c)d = 0,$

cannot be so expressed. The Teichmuller identity requires five functions to express it. We handle identities using more than one function by treating the identities as elements of a left module over a group ring.

Let R be a nonassociative algebra over a field F. Let

$$f_i : \prod_{j=1}^{n} R \longrightarrow R$$

be functions for $i = 1, 2, \ldots, r$. Let ρ be the group ring over F of the symmetric group S_n. An identity is an element

$$\Phi \in \prod_{i=1}^{r} \rho$$

such that the function

$$\sum_{i=1}^{r} (f_i)_{\Phi(i)}$$

is zero no matter what elements are used for its arguments. Φ
is called an identity of R based on the functions f_1, $f_2, \ldots, f_r$.
The direct sum

$$\prod_{i=1}^{r} \rho$$

is a left module over ρ, and the set of all identities based on
the functions f_1, $f_2, \ldots, f_r$ is a submodule. An identity Φ is a
consequence of $\Phi_1, \Phi_2, \ldots, \Phi_s$ if and only if Φ is in the submodule
generated by $\Phi_1, \Phi_2, \ldots, \Phi_s$.

If $P_1 \oplus P_2 \oplus \cdots \oplus P_r$ is an identity of R based on
$f_1, f_2, \ldots, f_r$, and P_k is invertible, then $P_k^{-1} P_1 \oplus P_k^{-1} P_2 \oplus \cdots$
$\oplus I_k \oplus \cdots \oplus P_k^{-1} P_r$ expresses the function f_k in terms of the
remaining r-1 functions. If $pp_\ell = 0$ for some $p \in \rho$, then $pp_1 \oplus pp_2$
$\oplus \cdots \oplus 0_\ell \oplus \cdots \oplus pp_r$ is an identity not involving f_ℓ. Using
these techniques, an identity involving several functions can be
reduced to an identity involving fewer functions.

4.   Identities Modulo An Additive Subgroup

Let R be any nonassociative algebra over a field F, H be any
additive subgroup of R (H not necessarily an ideal), and
$f: \prod_{i=1}^{n} R \to R$. We say $\Phi = \sum_{\pi \in S_n} \gamma_\pi \pi$ is an identity mod H of R
based on the function f if the range of $f_\Phi \subseteq H$. As before, the
set of all identities of R mod H based on the function f is a
left ideal of the group ring. A similar result applies to identities involving more than one function; there, we have a submodule mod H of identities.

5.   Applications

We shall apply the technique by classifying associator
dependent algebras. An associator dependent algebra is a nonassociative algebra over a field F that is assumed to satisfy
one or more identities of this form:

(2) $\alpha(a,b,c)+\beta(c,a,b)+\gamma(b,c,a)+\delta(b,a,c)+\varepsilon(a,c,b)+\eta(c,b,a)=0.$

The coefficients $\alpha,\beta,\gamma,\delta,\varepsilon,\eta$, are elements of F, and a,b,c are
elements of R. The expression (x,y,z) is called the associator
and is defined by $(x,y,z) = (xy)z-x(yz)$ for all $x,y,z \in R$. It
is natural to use the associator as the function upon which the
identities are based. The group ring ρ will then be the group
ring over F of the symmetric group S_3. The representation of ρ
is given in Table I. A non-trivial identity of form (2) is called
an associator dependent relation.

We now identify the distinct associator dependent algebras.
Two sets of associator dependent relations are called equivalent
if each set implies the other. This means they generate the same
left ideal in the group ring. The distinct associator dependent
algebras correspond to the distinct left ideals of the group
ring. If the characteristic of F is not 2 or 3, then the group
ring is isomorphic to:

$$F \oplus F \oplus F_{2\times 2}$$

where $F_{2\times 2}$ is the 2×2 matrices over F. Each left ideal L also
decomposes into $L = L_1 \oplus L_2 \oplus L_3$ where each L_i is a left ideal
of the summand which contains it. The classification of all
left ideals of the group ring corresponds to finding all choices
for the L_i. Since L_1 is a left ideal of a field, it is either 0
or the whole field. The same choices hold for L_2. L_3 may be
zero, all of $F_{2\times 2}$, or a proper left ideal of $F_{2\times 2}$. It will be
generated by a nonzero matrix of the form:

$$\begin{bmatrix} \lambda_1 & \lambda_2 \\ 0 & 0 \end{bmatrix}$$

Since every left ideal of ρ is the direct sum of minimal left ideals, every associator dependent algebra is obtained by assuming one or more minimal associator dependent relations. A minimal associator dependent relation is one which corresponds to a minimal left ideal. We will classify these minimal associator dependent relations. If a minimal left ideal L has a non-zero entry in the first summand, then $L = F \oplus 0 \oplus 0$, and L corresponds to the identity:

$$\sum_{\pi \varepsilon S_3} (a,b,c)_\pi = 0.$$

This is the linearized form of the identity $(x,x,x) = 0$ which is called third-power associativity. If L has a nonzero entry in the second component, then $L = 0 \oplus F \oplus 0$, and L corresponds to:

$$\sum_{\pi \varepsilon S_3} \text{sgn}\pi \, (a,b,c)_\pi = 0.$$

This is Lie admissibility. In associator form it is $(a,b,c) + (c,a,b) + (b,c,a) - (a,c,b) - (b,a,c) - (c,b,a) = 0$. It may also be expressed as:

(3) $[a,[b,c]] + [c,[a,b]] + [b,[c,a]] = 0$

where the commutator $[x,y]$ is defined by $[x,y] = xy-yx$. If L is obtained from a proper left ideal of $F_{2 \times 2}$, then L will be generated by some $0 \oplus 0 \oplus \begin{bmatrix} \lambda_1 & \lambda_2 \\ 0 & 0 \end{bmatrix}$. L corresponds to the identity:

$$\lambda_1 [(a,b,c) + (b,a,c) - (b,c,a) - (c,b,a)] + \lambda_2 [(a,c,b) - (c,a,b) + (b,c,a) - (b,a,c)] = 0.$$

It is easy to break up an identity into its minimal identities. The right alternative law, $I +(23)$, has representation $[2] \oplus [0] \oplus \begin{bmatrix} 1 & 1 \\ 1 & 1 \end{bmatrix}$. We can express an identity by any generator of the left ideal. Thus, the right alternative law could be expressed by $[1] \oplus [0] \oplus \begin{bmatrix} 1 & 1 \\ 0 & 0 \end{bmatrix}$ as well as by $[2] \oplus [0] \oplus \begin{bmatrix} 1 & 1 \\ 1 & 1 \end{bmatrix}$ We often write the matrices of the identities in row-canonical form to keep matters as simple as possible.

In Table II, we see that the right alternative law does not imply Lie admissibility. We also see that a $(-1,1)$ algebra is a Lie admissible right alternative algebra. The last three examples listed in Table II are maximal associator dependent algebras. If any other associator dependent relation were assumed which was not implied by those already assumed, then the algebra would be associative.

The larger the left ideal of identities, the easier it is to prove results. Simple $(-1,1)$, simple alternative, and simple (γ,δ) algebras have been classified. Simple right alternative algebras have not yet been classified.

The original classification of associator dependent rings was done in a paper by E. Kleinfeld, F. Kosier, J. M. Osborn, and D. Rodabaugh [7] without using group representation. The classification was done using group representation in a paper by D. Floyd and I. Hentzel [1].

We shall now present an example using group representations to prove identities in a $(-1,1)$ algebra. The results obtained here were done without using representation theory in [2]. The results were extended to (γ,δ) algebras using representation theory in [4]. If the reader wishes to delve further into the method, we recommend [4].

A $(-1,1)$ algebra is a nonassociative algebra which is defined by the following two identities.

TABLE I

THE REPRESENTATION OF S_3

I	[1]	$\oplus$	[1]	$\oplus$	$\begin{bmatrix} 1 & 0 \\ 0 & 1 \end{bmatrix}$	
(12)	[1]	$\oplus$	[-1]	$\oplus$	$\begin{bmatrix} 1 & 0 \\ -1 & -1 \end{bmatrix}$	
(13)	[1]	$\oplus$	[-1]	$\oplus$	$\begin{bmatrix} -1 & -1 \\ 0 & 1 \end{bmatrix}$	
(23)	[1]	$\oplus$	[-1]	$\oplus$	$\begin{bmatrix} 0 & 1 \\ 1 & 0 \end{bmatrix}$	
(123)	[1]	$\oplus$	[1]	$\oplus$	$\begin{bmatrix} -1 & -1 \\ 1 & 0 \end{bmatrix}$	
(132)	[1]	$\oplus$	[1]	$\oplus$	$\begin{bmatrix} 0 & 1 \\ -1 & -1 \end{bmatrix}$	

TABLE II

EXAMPLES OF ASSOCIATOR DEPENDENT ALBEGRAS

Identity	Name of Algebra
$[1] \oplus [0] \oplus \begin{bmatrix} 1 & 1 \\ 0 & 0 \end{bmatrix}$	Right Alternative
$[1] \oplus [1] \oplus \begin{bmatrix} 1 & 1 \\ 0 & 0 \end{bmatrix}$	$(-1,1)$
$[1] \oplus [0] \oplus \begin{bmatrix} 1 & 0 \\ 0 & 1 \end{bmatrix}$	Alternative
$[1] \oplus [1] \oplus \begin{bmatrix} 1 & \lambda \\ 0 & 0 \end{bmatrix}$	(γ,δ)
$\lambda \neq 0,\ 1/2,\ 2$	$(\gamma,\delta) \neq (\pm 1,0)$

$\bar{A}(a,b,c) = (a,b,c)+(a,c,b) \neq 0$ (right alternative law)

$\bar{B}(a,b,c) = (a,b,c)+(c,a,b)+(b,c,a) = 0$ (cyclic law)

We will show the important result that in a $(-1,1)$ algebra, $[a,[b,(c,d,e)]] = 0$ is an identity. This identity was instrumental in the proof that simple $(-1,1)$ algebras of characteristic $\neq 2$, $\neq 3$ are associative. We will be content with showing how group representation can be used to prove this identity.

In this proof we will use ρ = the group ring on S_4 over a field F of characteristic $\neq 2$, $\neq 3$. ρ is isomorphic to $M_1 \oplus M_2 \oplus M_3 \oplus M_4 \oplus M_5$ where the M_i are matrices over F. M_3 and M_4 are given in Table III. M_1 is the identity representation whose value is always 1. M_2 is the alternating representation whose value is the sign of the permutation. $M_5(\pi) = \text{sgn} \ (\pi)M_4(\pi)$.

We shall first prove that in a $(-1,1)$ algebra, the additive span of the associators, written (R,R,R), is actually an ideal of the algebra.

Lemma 1.   In a $(-1,1)$ algebra, the additive span of the associators is an ideal.

Proof.   We will base our proof on the function $R(R,R,R)$. The identities are calculated mod (R,R,R). We know $a\bar{A}(b,c,d) = 0$ and $a\bar{B}(b,c,d) = 0$. Using Teichmuller's equality (Eq. (1)) we have

$a(b,c,d)+a(c,b,d) \equiv -\bar{A}(a,b,c)d \equiv 0 \mod (R,R,R)$ and

$a(b,c,d)+c(a,b,d)+b(c,a,d) \equiv -\bar{B}(a,b,c)d \equiv 0 \mod (R,R,R)$.

We thus have four identities of R mod (R,R,R) based on the function $R(R,R,R)$. They are $I+(34)$, $I+(234)+(243)$, $I+(23)$, $I+(123)+(132)$. Their representations are given here.

Table III

THE REPRESENTATION OF S_4

π	π_3	π_4	π	π_3	π_4
(12)	$\begin{bmatrix} 1 & 0 \\ -1 & -1 \end{bmatrix}$	$\begin{bmatrix} -1 & 0 & 0 \\ -1 & 1 & 0 \\ -1 & 0 & 1 \end{bmatrix}$	I	$\begin{bmatrix} 1 & 0 \\ 0 & 1 \end{bmatrix}$	$\begin{bmatrix} 1 & 0 & 0 \\ 0 & 1 & 0 \\ 0 & 0 & 1 \end{bmatrix}$
(13)	$\begin{bmatrix} -1 & -1 \\ 0 & 1 \end{bmatrix}$	$\begin{bmatrix} 1 & -1 & 0 \\ 0 & -1 & 0 \\ 0 & -1 & 1 \end{bmatrix}$	(123)	$\begin{bmatrix} -1 & -1 \\ 1 & 0 \end{bmatrix}$	$\begin{bmatrix} -1 & 1 & 0 \\ -1 & 0 & 0 \\ -1 & 0 & 1 \end{bmatrix}$
(14)	$\begin{bmatrix} 0 & 1 \\ 1 & 0 \end{bmatrix}$	$\begin{bmatrix} 1 & 0 & -1 \\ 0 & 1 & -1 \\ 0 & 0 & -1 \end{bmatrix}$	(124)	$\begin{bmatrix} 0 & 1 \\ -1 & -1 \end{bmatrix}$	$\begin{bmatrix} -1 & 0 & 1 \\ -1 & 1 & 0 \\ -1 & 0 & 0 \end{bmatrix}$
(23)	$\begin{bmatrix} 0 & 1 \\ 1 & 0 \end{bmatrix}$	$\begin{bmatrix} 0 & 1 & 0 \\ 1 & 0 & 0 \\ 0 & 0 & 1 \end{bmatrix}$	(132)	$\begin{bmatrix} 0 & 1 \\ -1 & -1 \end{bmatrix}$	$\begin{bmatrix} 0 & -1 & 0 \\ 1 & -1 & 0 \\ 0 & -1 & 1 \end{bmatrix}$
(24)	$\begin{bmatrix} -1 & -1 \\ 0 & 1 \end{bmatrix}$	$\begin{bmatrix} 0 & 0 & 1 \\ 0 & 1 & 0 \\ 1 & 0 & 0 \end{bmatrix}$	(134)	$\begin{bmatrix} -1 & -1 \\ 1 & 0 \end{bmatrix}$	$\begin{bmatrix} 1 & -1 & 0 \\ 0 & -1 & 1 \\ 0 & -1 & 0 \end{bmatrix}$
(34)	$\begin{bmatrix} 1 & 0 \\ -1 & -1 \end{bmatrix}$	$\begin{bmatrix} 1 & 0 & 0 \\ 0 & 0 & 1 \\ 0 & 1 & 0 \end{bmatrix}$	(142)	$\begin{bmatrix} -1 & -1 \\ 1 & 0 \end{bmatrix}$	$\begin{bmatrix} 0 & 0 & -1 \\ 0 & 1 & -1 \\ 1 & 0 & -1 \end{bmatrix}$
(1234)	$\begin{bmatrix} -1 & -1 \\ 0 & 1 \end{bmatrix}$	$\begin{bmatrix} -1 & 1 & 0 \\ -1 & 0 & 1 \\ -1 & 0 & 0 \end{bmatrix}$	(143)	$\begin{bmatrix} 0 & 1 \\ -1 & -1 \end{bmatrix}$	$\begin{bmatrix} 1 & 0 & -1 \\ 0 & 0 & -1 \\ 0 & 1 & -1 \end{bmatrix}$
(1243)	$\begin{bmatrix} 0 & 1 \\ 1 & 0 \end{bmatrix}$	$\begin{bmatrix} -1 & 0 & 1 \\ -1 & 0 & 0 \\ -1 & 1 & 0 \end{bmatrix}$	(234)	$\begin{bmatrix} 0 & 1 \\ -1 & -1 \end{bmatrix}$	$\begin{bmatrix} 0 & 1 & 0 \\ 0 & 0 & 1 \\ 1 & 0 & 0 \end{bmatrix}$
(1324)	$\begin{bmatrix} 1 & 0 \\ -1 & -1 \end{bmatrix}$	$\begin{bmatrix} 0 & -1 & 1 \\ 1 & -1 & 0 \\ 0 & -1 & 0 \end{bmatrix}$	(243)	$\begin{bmatrix} -1 & -1 \\ 1 & 0 \end{bmatrix}$	$\begin{bmatrix} 0 & 0 & 1 \\ 1 & 0 & 0 \\ 0 & 1 & 0 \end{bmatrix}$
(1342)	$\begin{bmatrix} 0 & 1 \\ 1 & 0 \end{bmatrix}$	$\begin{bmatrix} 0 & -1 & 0 \\ 0 & -1 & 1 \\ 1 & -1 & 0 \end{bmatrix}$	(12)(34)	$\begin{bmatrix} 1 & 0 \\ 0 & 1 \end{bmatrix}$	$\begin{bmatrix} -1 & 0 & 0 \\ -1 & 0 & 1 \\ -1 & 1 & 0 \end{bmatrix}$
(1423)	$\begin{bmatrix} 1 & 0 \\ -1 & -1 \end{bmatrix}$	$\begin{bmatrix} 0 & 1 & -1 \\ 0 & 0 & -1 \\ 1 & 0 & -1 \end{bmatrix}$	(13)(24)	$\begin{bmatrix} 1 & 0 \\ 0 & 1 \end{bmatrix}$	$\begin{bmatrix} 0 & -1 & 1 \\ 0 & -1 & 0 \\ 1 & -1 & 0 \end{bmatrix}$
(1432)	$\begin{bmatrix} -1 & -1 \\ 0 & 1 \end{bmatrix}$	$\begin{bmatrix} 0 & 0 & -1 \\ 1 & 0 & -1 \\ 0 & 1 & -1 \end{bmatrix}$	(14)(23)	$\begin{bmatrix} 1 & 0 \\ 0 & 1 \end{bmatrix}$	$\begin{bmatrix} 0 & 1 & -1 \\ 1 & 0 & -1 \\ 0 & 0 & -1 \end{bmatrix}$

$$[2] \quad \oplus \quad [0] \quad \oplus \quad \begin{bmatrix} 2 & 0 \\ -1 & 0 \end{bmatrix} \oplus \begin{bmatrix} 2 & 0 & 0 \\ 0 & 1 & 1 \\ 0 & 1 & 1 \end{bmatrix} \oplus \begin{bmatrix} 0 & 0 & 0 \\ 0 & 1 & -1 \\ 0 & -1 & 1 \end{bmatrix}$$

$$[3] \quad \oplus \quad [3] \quad \oplus \quad \begin{bmatrix} 0 & 0 \\ 0 & 0 \end{bmatrix} \oplus \begin{bmatrix} 1 & 1 & 1 \\ 1 & 1 & 1 \\ 1 & 1 & 1 \end{bmatrix} \oplus \begin{bmatrix} 1 & 1 & 1 \\ 1 & 1 & 1 \\ 1 & 1 & 1 \end{bmatrix}$$

$$[2] \quad \oplus \quad [0] \quad \oplus \quad \begin{bmatrix} 1 & 1 \\ 1 & 1 \end{bmatrix} \oplus \begin{bmatrix} 1 & 1 & 0 \\ 1 & 1 & 0 \\ 0 & 0 & 2 \end{bmatrix} \oplus \begin{bmatrix} 1 & -1 & 0 \\ -1 & 1 & 0 \\ 0 & 0 & 0 \end{bmatrix}$$

$$[3] \quad \oplus \quad [3] \quad \oplus \quad \begin{bmatrix} 0 & 0 \\ 0 & 0 \end{bmatrix} \oplus \begin{bmatrix} 0 & 0 & 0 \\ 0 & 0 & 0 \\ -1 & -1 & 3 \end{bmatrix} \oplus \begin{bmatrix} 0 & 0 & 0 \\ 0 & 0 & 0 \\ -1 & -1 & 3 \end{bmatrix}$$

It is clear that

$$\Phi = [3] \quad \oplus \quad [3] \quad \oplus \quad \begin{bmatrix} 2 & 0 \\ 1 & 1 \end{bmatrix} \oplus \begin{bmatrix} 2 & 0 & 0 \\ 0 & 1 & 1 \\ 0 & 0 & 2 \end{bmatrix} \oplus \begin{bmatrix} 0 & 1 & -1 \\ 1 & 1 & 1 \\ -1 & -1 & 3 \end{bmatrix}$$

is in the left ideal generated by these identities. Since Φ is invertible, the identity element of the group ring I is also in the left ideal. This means

$$a(b,c,d)_I = a(b,c,d) \ \varepsilon \ (R,R,R)$$

for all $a,b,c,d \ \varepsilon R$. Therefore (R,R,R) is a left ideal. By Teichmuller's identity (1), it is also a right ideal.

The function $[R,(R,R,R)]$ is of considerable importance. There are many identities known involving its entries. Let us compute the left ideal of identities for it.

Lemma 2. Let R be a $(-1,1)$ algebra. The left ideal of identities of R based on the function $[R,(R,R,R)]$ contains

$$[1] \quad \oplus \quad [1] \quad \oplus \quad \begin{bmatrix} 1 & 0 \\ 0 & 1 \end{bmatrix} \oplus \begin{bmatrix} 1 & 0 & 0 \\ 0 & 1 & 0 \\ 0 & 0 & 1 \end{bmatrix} \oplus \begin{bmatrix} 1 & 1 & 1 \\ 0 & -1 & 1 \\ 0 & 0 & 0 \end{bmatrix}$$

Proof. $[a,\bar{A}(b,c,d)] = 0$. $[a,\bar{\bar{B}}(b,c,d)] = 0$.   By the cyclic
law we also have

$\bar{C}(a,b,c,d)=[a,(b,c,d)]-[d,(a,b,c)]+[c,(d,a,b)]-[b,(c,d,a)]=0$.

These identities are I+(34), I+(234)+(243), I-(1234)+(13)(24)-
(1432).   They have these representations.

$$[2] \;\oplus\; [0] \;\oplus\; \begin{bmatrix} 2 & 0 \\ -1 & 0 \end{bmatrix} \;\oplus\; \begin{bmatrix} 2 & 0 & 0 \\ 0 & 1 & 1 \\ 0 & 1 & 1 \end{bmatrix} \;\oplus\; \begin{bmatrix} 0 & 0 & 0 \\ 0 & 1 & -1 \\ 0 & -1 & 1 \end{bmatrix}$$

$$[3] \;\oplus\; [3] \;\oplus\; \begin{bmatrix} 0 & 0 \\ 0 & 0 \end{bmatrix} \;\oplus\; \begin{bmatrix} 1 & 1 & 1 \\ 1 & 1 & 1 \\ 1 & 1 & 1 \end{bmatrix} \;\oplus\; \begin{bmatrix} 1 & 1 & 1 \\ 1 & 1 & 1 \\ 1 & 1 & 1 \end{bmatrix}$$

$$[0] \;\oplus\; [4] \;\oplus\; \begin{bmatrix} 4 & 2 \\ 0 & 0 \end{bmatrix} \;\oplus\; \begin{bmatrix} 2 & -2 & 2 \\ 0 & 0 & 0 \\ 2 & -2 & 2 \end{bmatrix} \;\oplus\; \begin{bmatrix} 0 & 0 & 0 \\ 0 & 0 & 0 \\ 0 & 0 & 0 \end{bmatrix}$$

We notice now that for each summand, the rows of the proposed
identity are in the space spanned by the rows of the identities.
This completes the proof.

Lemma 3.   In a (-1,1) algebra R, $[a,(b,b,a)] = 0$ for all a,b ε R.

Proof.   We wish to know if $[I+(14)][I+(23)]$ is in the left
ideal of identities.   By Lemma 2, we need only to check the
fifth component.   The fifth component of $[I+(14)][I+(23)]$ is
$0_{3\times3}$.   Since $[I+(14)][I+(23)]$ is in the left ideal generated
by I+(34), I+(234)+(243), and I-(1234)+(13)(24)-(1432), we have
shown that $[a,(b,b,a)] = 0$ is a consequence of $\bar{A}, \bar{B}$, and $\bar{C}$.

The left ideal of identities of $[R,(R,R,R)]$ is maximal.   This
means there are numerous results like Lemma 3 that could be proved.
The result is that a person continues to produce more and more of
these identities and feels he is getting somewhere, while what he
actually is doing is writing equivalent forms of identities.   From
the group representation approach, we know that $[R,(R,R,R)]$ is
almost zero.   We also know that if we are to prove it is zero,
we have to use a new identity not implied by the three we already

used. This lets us stop applying and reapplying our original
three identities. We can look for a new identity without the
nagging suspicion that, had we applied our three identities
in yet one more way, we would have gotten $[a,(b,c,d)] = 0$.

There are many ways we could proceed. We could include
$[a,(b,c,d)]$ inside an associator as $([a,(b,c,d)],e,f)$ or inside
a commutator as $[x,[a,(b,c,d)]]$ and try to get a new identity
which holds in this special situation. If we find one, we would
know that $[a,(b,c,d)]$ is in the left nucleus or that $[a,(b,c,d)]$
commutes with everything. Both of these approaches would probably
be used. They require the representation of S_6 and S_5. We shall
proceed in a different way so that the method will not be lost in
the computation.

We shall need this lemma shortly.

Lemma 4.   Let R be a $(-1,1)$ algebra. The left ideal of identities of R based on the function $R(R,R,R)$ contains

$$[1] \oplus [1] \begin{bmatrix} 1 & 0 \\ 0 & 0 \end{bmatrix} \oplus \begin{bmatrix} 1 & 0 & 0 \\ 0 & 1 & 1 \\ 0 & 0 & 0 \end{bmatrix} \oplus \begin{bmatrix} 1 & 1 & 1 \\ 0 & -1 & 1 \\ 0 & 0 & 0 \end{bmatrix}$$

Proof.   The left hand sides of $a\bar{A}(b,c,d) = 0$ and $a\bar{B}(b,c,d) = 0$
are, respectively, $I+(34)$ and $I+(234)+(243)$. The matrix representation of these two elements is

$$[2] \oplus [0] \oplus \begin{bmatrix} 2 & 0 \\ -1 & 0 \end{bmatrix} \oplus \begin{bmatrix} 2 & 0 & 0 \\ 0 & 1 & 1 \\ 0 & 1 & 1 \end{bmatrix} \oplus \begin{bmatrix} 0 & 0 & 0 \\ 0 & 1 & -1 \\ 0 & -1 & 1 \end{bmatrix}$$

and

$$[3] \oplus [3] \oplus \begin{bmatrix} 0 & 0 \\ 0 & 0 \end{bmatrix} \oplus \begin{bmatrix} 1 & 1 & 1 \\ 1 & 1 & 1 \\ 1 & 1 & 1 \end{bmatrix} \oplus \begin{bmatrix} 1 & 1 & 1 \\ 1 & 1 & 1 \\ 1 & 1 & 1 \end{bmatrix}$$

The result follows.

When R is a (-1,1) algebra, we let

$$U = \{u \in R \mid [x,u] = 0 \text{ for all } x \in R\}.$$

U is an additive subgroup of R. Maneri [6, Eq. (11)] proved that for all x and for all d, $(x,x,[x,d]) \in U$. A key lemma in determining the structure of (-1,1) algebras was strengthening this result to

$$(x,x,[c,d]) \in U$$

for all x,c,d in R. If we examine identities based on (R,R,RR) mod U, we have by Maneri's result $[I+(123)+(132)][I+(12)][I-(34)]$ is an element of the left ideal of identities mod U. We wish to show $[I+(12)][I-(34)]$ is also in the left ideal of identities mod U. The representations of these two elements are listed below.

$(x,x,[x,d])$	$(x,x,[c,d])$
$[I+(123)+(132)][I+(12)][I-(34)]$	$[I+(12)][I-(34)]$

$$[0] \qquad\qquad\qquad [0]$$

$$[0] \qquad\qquad\qquad [0]$$

$$\begin{bmatrix} 0 & 0 \\ 0 & 0 \end{bmatrix} \qquad\qquad \begin{bmatrix} 0 & 0 \\ 0 & 0 \end{bmatrix}$$

$$\begin{bmatrix} 0 & 0 & 0 \\ 0 & 0 & 0 \\ 0 & -8 & -8 \end{bmatrix} \qquad \begin{bmatrix} 0 & 0 & 0 \\ 0 & 2 & -2 \\ 0 & -2 & 2 \end{bmatrix}$$

$$\begin{bmatrix} 0 & 0 & 0 \\ 0 & 0 & 0 \\ 0 & 0 & 0 \end{bmatrix} \qquad \begin{bmatrix} 4 & 0 & 0 \\ 2 & 0 & 0 \\ 2 & 0 & 0 \end{bmatrix}$$

In summands 1,2,3 and 4, it is clear that the left ideal generated by the representation of the identity $(x,x,[x,a])$ contains the representation of $(x,x,[c,d])$. The proof will be complete if we can show that in the fifth summand

$$\begin{bmatrix} 1 & 0 & 0 \\ 0 & 0 & 0 \\ 0 & 0 & 0 \end{bmatrix}$$

is an identity mod U. To show this requires us to use the information in Lemma 2 and Lemma 4 to tell us about (R,R,RR). This will illustrate how to use more than one function.

The functions we will deal with are (R,R,RR), R(R,R,R), and [R,(R,R,R)]. ρ will be the group ring on S_4 over F. Rather than number our functions, we shall separate them with " $\oplus$ ". Thus, where f,f', and f" are functions, we will write f $\oplus$ f' $\oplus$ f", and it is understood that an element g $\oplus$ g' $\oplus$ g" in $\rho \oplus \rho \oplus \rho$ will apply g to f, g' to f' and g" to f". In the representation of $\rho \oplus \rho \oplus \rho$ each summand will have five corresponding parts. $\rho \oplus \rho \oplus \rho$ is represented by:

$$(4) \qquad \begin{bmatrix} M_1 \\ M_2 \\ M_3 \\ M_4 \\ M_5 \end{bmatrix} \oplus \begin{bmatrix} M'_1 \\ M'_2 \\ M'_3 \\ M'_4 \\ M'_5 \end{bmatrix} \oplus \begin{bmatrix} M''_1 \\ M''_2 \\ M''_3 \\ M''_4 \\ M''_5 \end{bmatrix}$$

This display is crucial to efficient dealing with identities. The set L of all elements $\ell \oplus \ell' \oplus \ell''$ of $\rho \oplus \rho \oplus \rho$ such that for any arguments a,b,c,d ε R, f_ℓ(a,b,c,d)+f'$_{\ell'}$(a,b,c,d)+f''$_{\ell''}$(a,b,c,d)=0, forms a submodule of $\rho \oplus \rho \oplus \rho$ over ρ. Since it is always possible to find an element of ρ which multiplies a particular row of a matrix by 1 and all other rows by 0, we can say that each row of the display is an identity of R. If we stay within the rows of the same representation, we can perform any row operation and still maintain an identity of R. It is important to perform the same row operation on the entire row. In our example, the row extends across three separate matrices. The minimal identities correspond to rows of the display. The next larger unit of identities corresponds to rows which come from the same representation. We shall often deal with rows corresponding to the same representation and indicate it in this way:

In Block #, $M \oplus M' \oplus M''$ is an identity.

Here, M, M', and M'' are matrices, # is 1, 2, 3, 4, or 5. $M \oplus M' \oplus M''$ is to be considered as a Block of a complete display as given in (4) where all the non-listed Blocks are zero. By the above discussion, an element with a display given in (4) is an identity if and only if, in Block i, $M_i \oplus M'_i \oplus M''_i$ is an identity for each i.

Rewriting the Teichmuller identity (1), using $\bar{A}$ and $\bar{B}$, we get

$$\{(a,b,cd)-(c,d,ab)+(d,c,ab)+(a,d,bc)\}+$$
$$\{-a(b,c,d)-d(a,b,c)\}+[d,(a,b,c)] = 0.$$

Writing this identity using group algebra notation based on the functions

$$(R,R,RR) \oplus R(R,R,R) \oplus [R,(R,R,R)],$$

the element of $\rho \oplus \rho \oplus \rho$ is given as

$$\{I-(13)(24)+(1324)+(234)\} \oplus \{-I-(1234)\} \oplus (1234).$$

In Table IV we have expressed this element of $\rho \oplus \rho \oplus \rho$ using the matrix representation in the manner of (4).

Lemma 4 gives us the identities for the function R(R,R,R). If ℓ' is the element listed in Lemma 4, then $0 \oplus \ell' \oplus 0$ will be an identity for $(R,R,RR) \oplus R(R,R,R) \oplus [R,(R,R,R)]$. Similarly if ℓ'' is the identity of $[R,(R,R,R)]$ given in Lemma 2, then $0 \oplus 0 \oplus \ell''$ is an identity for $(R,R,RR) \oplus R(R,R,R) \oplus [R,(R,R,R)]$. It should be noted that the representations of ℓ' and ℓ'' in Lemma 4 and Lemma 2 are written horizontally and should be rewritten vertically to conform with usage in Table IV. From Lemma 2 we deduce that in Block 4, $0 \oplus 0 \oplus I_{3 \times 3}$ is an identity.

Table IV

THE IDENTITIES OF A (-1,1) ALGEBRA

	(R,R,RR) ⊕	R(R,R,R) ⊕	[R,(R,R,R)]
Block 1	[2]	*	*
Block 2	[0]	*	*
Block 3	$\begin{bmatrix} 1 & 1 \\ -2 & -2 \end{bmatrix}$	$\begin{bmatrix} 0 & 1 \\ 0 & -2 \end{bmatrix}$	*
Block 4	$\begin{bmatrix} 1 & 1 & 0 \\ 1 & 1 & 1 \\ 0 & 0 & 1 \end{bmatrix}$	$\begin{bmatrix} 0 & -1 & 0 \\ 1 & -1 & -1 \\ 1 & 0 & -1 \end{bmatrix}$	*
Block 5	$\begin{bmatrix} 1 & 3 & -2 \\ -1 & 3 & 1 \\ 0 & 2 & 1 \end{bmatrix}$	$\begin{bmatrix} -2 & 1 & 0 \\ -1 & -1 & 1 \\ -1 & 0 & -1 \end{bmatrix}$	$\begin{bmatrix} 1 & -1 & 0 \\ 1 & 0 & -1 \\ 1 & 0 & 0 \end{bmatrix}$

This means that in Block 4 any expression of the form $0 \oplus 0 \oplus M_{3 \times 3}$ is an identity. This says that the entry of Block 4 under $[R,(R,R,R)]$ can be ignored; no matter what it is, it maps to zero. This is the meaning of the asterisks appearing in Table IV. Here is an example of what we can learn from Table IV. From Block 1 we deduce that $[2] \oplus [0] \oplus [0]$ is an identity; thus $(x,x,x^2) = 0$.

Using Lemma 2 and Lemma 4 we find that in Block 5

$$
(5) \qquad
\begin{bmatrix} 1 & 3 & -2 \\ -1 & 3 & 1 \\ 0 & 2 & 1 \end{bmatrix}
\oplus
\begin{bmatrix} 0 & 0 & 5 \\ 0 & 0 & 2 \\ 0 & 0 & 1 \end{bmatrix}
\oplus
\begin{bmatrix} 0 & 0 & -3 \\ 0 & 0 & -3 \\ 0 & 0 & -2 \end{bmatrix}
$$

is an identity. Multiplying across by $\begin{bmatrix} -1 & 7 & -9 \\ 0 & 0 & 0 \\ 0 & 0 & 0 \end{bmatrix}$ produces an

identity of the form $\ell \oplus 0 \oplus 0$ where $\ell = \begin{bmatrix} -8 & 0 & 0 \\ 0 & 0 & 0 \\ 0 & 0 & 0 \end{bmatrix}$

We state this result in the following lemma.

Lemma 5.   In Block 5

$$
\begin{bmatrix} 1 & 0 & 0 \\ 0 & 0 & 0 \\ 0 & 0 & 0 \end{bmatrix}
\oplus 0 \oplus 0
$$

is an identity of a $(-1,1)$ algebra based on $(R,R,RR) \oplus R(R,R,R) \oplus [R,(R,R,R)]$.

Lemma 5 says that $(x,x,[x,d]) \; \varepsilon \; U$ for all $x,d \; \varepsilon \; R$ implies $(x,x,[c,d]) \; \varepsilon \; U$ for all $x,c,d \; \varepsilon \; R$. This completes the proof of the following theorem.

Theorem 1.   For all c,d,x,z in a $(-1,1)$ algebra of characteristic $\neq 2, \neq 3$, $[z,(x,x,[c,d])] = 0$.

A $(-1,1)$ algebra is Lie admissible. By (3), we have $[a,[b,c]]+[c,[a,b]]+[b,[c,a]]=0$ for all elements a,b,c in R. We apply this identity to $[c,[d,(x,x,z)]]$.

$$[c,[d,(x,x,z)]] = -[(x,x,z),[c,d]]-[d,[(x,x,z),c]]$$
$$=-[z,(x,x,[c,d])]+[d,[c,(x,x,z)]]$$
$$= [d,[c,(x,x,z)]]$$

by Lemma 3 and Theorem 1. Using the 1,2,5 positions we have $[c,[d,(x,x,z)]] = [d,[c,(x,x,z)]] = -[d,[z,(x,x,c)]]$. Here a cyclic permutation changes the sign; we have, after performing three cyclic permutations, $[c,[d,(x,x,z)]]=-[c,[d,(x,x,z)]]$. We have shown $[c,[d,(x,x,z)]] = 0$ for all c,d,x,z in R.

<u>Theorem 2.</u>   In a (-1,1) algebra R of characteristic $\neq 2, \neq 3$, $[R,[R,(R,R,R)]] = 0$.

<u>Proof.</u>   We have shown that $[c,[d,(x,x,z)]] = 0$ for all c,d,x,z ε R. It remains to show that $(R,R,R) \subseteq A$ = the additive span of $\{(x,x,z)\,|\,x,z \varepsilon R\}$. By Lie admissibility
$$0 = \underset{\pi\varepsilon S_3}{\Sigma}\ \mathrm{sgn}\pi(a,b,c)_\pi \equiv 6\ (a,b,c)\ \mathrm{mod}\ A.$$

This completes the proof of the theorem. Let us now analyze the types of elements that we know are in U. We know two results: $(x,x,[c,d]) \varepsilon U$ and $[a,(b,c,d)] \varepsilon U$ for all a,b,c,d,x ε R. We ask, which is stronger? It is clear that since (R,R,R) is an ideal, we can express $[a,(b,c,d)]$ as
$$\underset{\pi\varepsilon S_4}{\Sigma}\ \gamma_\pi\ (a,b,cd)_\pi\ .$$
It is easy to do this using (5). Multiplying through (5) by
$\begin{bmatrix} 0 & 0 & 0 \\ 0 & 0 & 0 \\ 0 & 1 & -2 \end{bmatrix}$ produces:

in Block 5 $\begin{bmatrix} 0 & 0 & 0 \\ 0 & 0 & 0 \\ -1 & -1 & -1 \end{bmatrix} \oplus 0 \oplus \begin{bmatrix} 0 & 0 & 0 \\ 0 & 0 & 0 \\ 0 & 0 & 1 \end{bmatrix}$ is an identity.

By Lemma 2 we get

in Block 5 $\begin{bmatrix} 0 & 0 & 0 \\ 0 & 0 & 0 \\ -1 & -1 & -1 \end{bmatrix} \oplus 0 \oplus \begin{bmatrix} 1 & 1 & 1 \\ 0 & -1 & 1 \\ 0 & 0 & 1 \end{bmatrix}$ is an identity.

The right hand matrix is invertible. Multiplying by its inverse produces:

$$\text{in Block 5} \begin{bmatrix} 2 & 2 & 2 \\ -1 & -1 & -1 \\ -1 & -1 & -1 \end{bmatrix} \oplus 0_{3\times3} \oplus I_{3\times3} \qquad \text{is an identity.}$$

By Lemma 2 $[a,(b,c,d)] + (a,b,cd)_g = 0$ where the representation of g is:

$$[0] \oplus [0] \oplus \begin{bmatrix} 0 & 0 \\ 0 & 0 \end{bmatrix} \oplus \begin{bmatrix} 0 & 0 & 0 \\ 0 & 0 & 0 \\ 0 & 0 & 0 \end{bmatrix} \oplus \begin{bmatrix} 2 & 2 & 2 \\ -1 & -1 & -1 \\ -1 & -1 & -1 \end{bmatrix}$$

By Theorem 2, $[R,(R,R,R)] \subseteq U$. Therefore $(a,b,cd)_g \in U$ for all a,b,c,d in R. This means that g is an identity mod U based on (R,R,RR).

The result that $[R,(R,R,R)] \subseteq U$ implies that based on (R,R,RR)

$$[0] \oplus [0] \oplus \begin{bmatrix} 0 & 0 \\ 0 & 0 \end{bmatrix} \oplus \begin{bmatrix} 0 & 0 & 0 \\ 0 & 0 & 0 \\ 0 & 0 & 0 \end{bmatrix} \oplus \begin{bmatrix} 1 & 1 & 1 \\ 0 & 0 & 0 \\ 0 & 0 & 0 \end{bmatrix}$$

is an identity mod U.

The identity $(x,x,[c,d]) \in U$ of Theorem 1 implies that based on (R,R,RR)

$$[0] \oplus [0] \oplus \begin{bmatrix} 0 & 0 \\ 0 & 0 \end{bmatrix} \oplus \begin{bmatrix} 0 & 1 & -1 \\ 0 & 0 & 0 \\ 0 & 0 & 0 \end{bmatrix} \oplus \begin{bmatrix} 1 & 0 & 0 \\ 0 & 0 & 0 \\ 0 & 0 & 0 \end{bmatrix}$$

is an identity mod U. It is strange that neither one is stronger. They, in fact, are disjoint. Together they imply $(a,b,[c,d]) - (c,d,[a,b]) \in U$ for all a,b,c,d $\in$ R. The representation of $[I-(13)(24)][I-(34)]$ is

$$[0] \oplus [0] \oplus \begin{bmatrix} 0 & 0 \\ 0 & 0 \end{bmatrix} \oplus \begin{bmatrix} 0 & 2 & -2 \\ 0 & 2 & -2 \\ 0 & 0 & 0 \end{bmatrix} \oplus \begin{bmatrix} 2 & 0 & 0 \\ 0 & 2 & 2 \\ -2 & 2 & 2 \end{bmatrix}$$

6.   Representation

Permutations apply to *positions*, not elements. Thus, if $\pi = (123)$, then $(a,b,c)_\pi = (c,a,b)$. We caution that $(x_1,x_2,x_3)_\pi = (x_3,x_1,x_2)$; $(x_1,x_2,x_3)_\pi$ is *not* equal to

$(x_{1\pi}, x_{2\pi}, x_{3\pi}) = (x_2, x_3, x_1)$. We multiply from left to right.
If $\sigma = (12)$, then $(a,b,c)_{\pi\sigma} = ((a,b,c)_\pi)_\sigma = (c,a,b)_\sigma = (a,c,b)$.

Table V contains representations of a sample of the elements of the group ring on S_4. They save time when a particular identity has to be computed time and time again.

In the listing we associate each identity with its linearized form. This does not mean that the group representation approach will apply only to linear functions. It simply reflects the fact that many functions are linear and the linearized form of an expression is useful. Thus, (a,x,x,x) is represented by $I+(234)+(243)+(23)+(34)+(24)$ or, equivalently, by $[I+(234)+(243)][I+(23)]$. If a non-listed representation is desired, it often can be derived from one on the list by multiplying by the representation of the appropriate permutation. If the representation for (a,c,c,b) is needed, multiply the representation of (a,b,c,c) on the right by the representation of the permutation (24).

Representations 9, 10, 11 and 12 of Table V are used when we have an identity whose matrix representation is known and we want to express it as a linear combination of permutations.

Table VI gives the Teichmuller identity for a right alternative ring. The function $M(x,y,z)$ is defined by $M(x,y,z) = (x,y,z) + (y,x,z)$.

Table VII combines the Teichmuller identity of Table VI, the right alternative law, and $M(ab,c,d) + M(ab,d,c) + (c,d,ab) + (d,c,ab) = 0$. The resulting generators of the left module of identities are then reduced to row canonical form. It represents all the degree four identities of right alternative rings.

Table V

USEFUL REPRESENTATIONS OF DEGREE FOUR

1.   (a,a,a,a) $\sum\limits_{\pi \varepsilon S_4} \pi$

$$[24] \oplus [0] \oplus \begin{bmatrix} 0 & 0 \\ 0 & 0 \end{bmatrix} \oplus \begin{bmatrix} 0 & 0 & 0 \\ 0 & 0 & 0 \\ 0 & 0 & 0 \end{bmatrix} \oplus \begin{bmatrix} 0 & 0 & 0 \\ 0 & 0 & 0 \\ 0 & 0 & 0 \end{bmatrix}$$

2.   (a,a,a,d)   [I+(123)+(132)][I+(12)]

$$[6] \oplus [0] \oplus \begin{bmatrix} 0 & 0 \\ 0 & 0 \end{bmatrix} \oplus \begin{bmatrix} 0 & 0 & 0 \\ 0 & 0 & 0 \\ -2 & -2 & 6 \end{bmatrix} \oplus \begin{bmatrix} 0 & 0 & 0 \\ 0 & 0 & 0 \\ 0 & 0 & 0 \end{bmatrix}$$

3.   (a,a,a,d)-(a,a,d,a)   [I+(123)+(132)][I+(12)][I-(34)]

$$[0] \oplus [0] \oplus \begin{bmatrix} 0 & 0 \\ 0 & 0 \end{bmatrix} \oplus \begin{bmatrix} 0 & 0 & 0 \\ 0 & 0 & 0 \\ 0 & -8 & 8 \end{bmatrix} \oplus \begin{bmatrix} 0 & 0 & 0 \\ 0 & 0 & 0 \\ 0 & 0 & 0 \end{bmatrix}$$

4.   (a,a,c,d)-(a,a,d,c)   [I+(12)][I-(34)]

$$[0] \oplus [0] \oplus \begin{bmatrix} 0 & 0 \\ 0 & 0 \end{bmatrix} \oplus \begin{bmatrix} 0 & 0 & 0 \\ 0 & 2 & -2 \\ 0 & -2 & 2 \end{bmatrix} \oplus \begin{bmatrix} 4 & 0 & 0 \\ 2 & 0 & 0 \\ 2 & 0 & 0 \end{bmatrix}$$

5.   (a,b,a,b)-(a,b,b,a)   [I+(13)][I+(24)][I-(34)]

$$[0] \oplus [0] \oplus \begin{bmatrix} -2 & -4 \\ 4 & 8 \end{bmatrix} \oplus \begin{bmatrix} 0 & -4 & 4 \\ 0 & 0 & 0 \\ 0 & -4 & 4 \end{bmatrix} \oplus \begin{bmatrix} 0 & 0 & 0 \\ 0 & 0 & 0 \\ 0 & 0 & 0 \end{bmatrix}$$

6.   (a,a,c,c)   [I+(12)][I+(34)]

$$[4] \oplus [0] \oplus \begin{bmatrix} 4 & 0 \\ -2 & 0 \end{bmatrix} \oplus \begin{bmatrix} 0 & 0 & 0 \\ -2 & 2 & 2 \\ -2 & 2 & 2 \end{bmatrix} \oplus \begin{bmatrix} 0 & 0 & 0 \\ 0 & 0 & 0 \\ 0 & 0 & 0 \end{bmatrix}$$

7.   (a,b,c,d)+(a,d,b,c)+(a,c,d,b)   I+(234)+(243)

$$[3] \oplus [3] \oplus \begin{bmatrix} 0 & 0 \\ 0 & 0 \end{bmatrix} \oplus \begin{bmatrix} 1 & 1 & 1 \\ 1 & 1 & 1 \\ 1 & 1 & 1 \end{bmatrix} \oplus \begin{bmatrix} 1 & 1 & 1 \\ 1 & 1 & 1 \\ 1 & 1 & 1 \end{bmatrix}$$

Table V

(Continued)

8.   (a,b,c,c)   I+(34)

$$[2] \oplus [0] \oplus \begin{bmatrix} 2 & 0 \\ -1 & 0 \end{bmatrix} \oplus \begin{bmatrix} 2 & 0 & 0 \\ 0 & 1 & 1 \\ 0 & 1 & 1 \end{bmatrix} \oplus \begin{bmatrix} 0 & 0 & 0 \\ 0 & 1 & -1 \\ 0 & -1 & 1 \end{bmatrix}$$

9.   Sum of all two-cycles

$$[6] \oplus [-6] \oplus \begin{bmatrix} 0 & 0 \\ 0 & 0 \end{bmatrix} \oplus \begin{bmatrix} 2 & 0 & 0 \\ 0 & 2 & 0 \\ 0 & 0 & 2 \end{bmatrix} \oplus \begin{bmatrix} -2 & 0 & 0 \\ 0 & -2 & 0 \\ 0 & 0 & -2 \end{bmatrix}$$

10.   Sum of all three-cycles

$$[8] \oplus [8] \oplus \begin{bmatrix} -4 & 0 \\ 0 & -4 \end{bmatrix} \oplus \begin{bmatrix} 0 & 0 & 0 \\ 0 & 0 & 0 \\ 0 & 0 & 0 \end{bmatrix} \oplus \begin{bmatrix} 0 & 0 & 0 \\ 0 & 0 & 0 \\ 0 & 0 & 0 \end{bmatrix}$$

11.   Sum of all four-cycles

$$[6] \oplus [-6] \oplus \begin{bmatrix} 0 & 0 \\ 0 & 0 \end{bmatrix} \oplus \begin{bmatrix} -2 & 0 & 0 \\ 0 & -2 & 0 \\ 0 & 0 & -2 \end{bmatrix} \oplus \begin{bmatrix} 2 & 0 & 0 \\ 0 & 2 & 0 \\ 0 & 0 & 2 \end{bmatrix}$$

12.   Sum of all two-two cycles

$$[3] \oplus [3] \oplus \begin{bmatrix} 3 & 0 \\ 0 & 3 \end{bmatrix} \oplus \begin{bmatrix} -1 & 0 & 0 \\ 0 & -1 & 0 \\ 0 & 0 & -1 \end{bmatrix} \oplus \begin{bmatrix} -1 & 0 & 0 \\ 0 & -1 & 0 \\ 0 & 0 & -1 \end{bmatrix}$$

Table VI

THE TEICHMULLER IDENTITY IN A RIGHT ALTERNATIVE ALGEBRA

M(RR,R,R)	⊕	(R,R,RR)	⊕	R(R,R,R)	⊕	[R,(R,R,R)]
I	⊕	I+(234)+(13)(24)	⊕	-I-(1234)	⊕	(1234)

[1]	[3]	[-2]	[1]
[1]	[3]	[0]	[-1]

$$\begin{bmatrix} 1 & 0 \\ 0 & 1 \end{bmatrix} \qquad \begin{bmatrix} 2 & 1 \\ -1 & 1 \end{bmatrix} \qquad \begin{bmatrix} 0 & 1 \\ 0 & -2 \end{bmatrix} \qquad \begin{bmatrix} -1 & -1 \\ 0 & 1 \end{bmatrix}$$

$$\begin{bmatrix} 1 & 0 & 0 \\ 0 & 1 & 0 \\ 0 & 0 & 1 \end{bmatrix} \quad \begin{bmatrix} 1 & 0 & 1 \\ 0 & 0 & 1 \\ 2 & -1 & 1 \end{bmatrix} \quad \begin{bmatrix} 0 & -1 & 0 \\ 1 & -1 & -1 \\ 1 & 0 & -1 \end{bmatrix} \quad \begin{bmatrix} -1 & 1 & 0 \\ -1 & 0 & 1 \\ -1 & 0 & 0 \end{bmatrix}$$

$$\begin{bmatrix} 1 & 0 & 0 \\ 0 & 1 & 0 \\ 0 & 0 & 1 \end{bmatrix} \quad \begin{bmatrix} 1 & 0 & 1 \\ 0 & 0 & 1 \\ 2 & -1 & 1 \end{bmatrix} \quad \begin{bmatrix} -2 & 1 & 0 \\ -1 & -1 & 1 \\ -1 & 0 & -1 \end{bmatrix} \quad \begin{bmatrix} 1 & -1 & 0 \\ 1 & 0 & -1 \\ 1 & 0 & 0 \end{bmatrix}$$

Table VII

RIGHT ALTERNATIVE ALGEBRAS

M(RR,R,R) $\oplus$ (R,R,RR) $\oplus$ R(R,R,R) $\oplus$ [R,(R,R,R)]

*   *   *   *

[1]   [3]   [0]   [-1]

$$
\begin{bmatrix} 1 & 0 \\ 0 & 1 \\ 0 & 0 \\ 0 & 0 \\ 0 & 0 \end{bmatrix}
\quad
\begin{bmatrix} 0 & -1 \\ 0 & 2 \\ 1 & 1 \\ 0 & 0 \\ 0 & 0 \end{bmatrix}
\quad
\begin{bmatrix} 0 & -1 \\ 0 & -1 \\ 0 & 1 \\ 1 & 0 \\ 0 & 0 \end{bmatrix}
\quad
\begin{bmatrix} 0 & 1 \\ 0 & 0 \\ 0 & -1 \\ 0 & 0 \\ 1 & 0 \end{bmatrix}
$$

$$
\begin{bmatrix} 1 & 0 & 0 \\ 0 & 1 & 0 \\ 0 & 0 & 1 \\ 0 & 0 & 0 \\ 0 & 0 & 0 \\ 0 & 0 & 0 \\ 0 & 0 & 0 \\ 0 & 0 & 0 \\ 0 & 0 & 0 \end{bmatrix}
\quad
\begin{bmatrix} 0 & -1 & 0 \\ 0 & 0 & 0 \\ 0 & -3 & 0 \\ 1 & 1 & 0 \\ 0 & 0 & 1 \\ 0 & 0 & 0 \\ 0 & 0 & 0 \\ 0 & 0 & 0 \\ 0 & 0 & 0 \end{bmatrix}
\quad
\begin{bmatrix} 0 & 0 & 1 \\ 0 & 0 & 1 \\ 0 & 0 & -2 \\ 0 & 0 & 1 \\ 0 & 0 & -1 \\ 1 & 0 & 0 \\ 0 & 1 & 1 \\ 0 & 0 & 0 \\ 0 & 0 & 0 \end{bmatrix}
\quad
\begin{bmatrix} 0 & 0 & -1 \\ 0 & 0 & 0 \\ 0 & 0 & 1 \\ 0 & 0 & -1 \\ 0 & 0 & 1 \\ 0 & 0 & 0 \\ 0 & 0 & 0 \\ 1 & 0 & 0 \\ 0 & 1 & 1 \end{bmatrix}
$$

$$
\begin{bmatrix} 1 & 0 & 0 \\ 0 & 1 & 0 \\ 0 & 0 & 1 \\ 0 & 0 & 0 \\ 0 & 0 & 0 \\ 0 & 0 & 0 \end{bmatrix}
\quad
\begin{bmatrix} 0 & 1 & 1 \\ 0 & 0 & 1 \\ 0 & 1 & 1 \\ 1 & -1 & 0 \\ 0 & 0 & 0 \\ 0 & 0 & 0 \end{bmatrix}
\quad
\begin{bmatrix} -2 & 0 & 2 \\ -1 & 0 & 0 \\ -1 & 0 & 1 \\ 0 & 0 & -1 \\ 0 & 1 & -1 \\ 0 & 0 & 0 \end{bmatrix}
\quad
\begin{bmatrix} 1 & 0 & -2 \\ 1 & 0 & -1 \\ 1 & 0 & -2 \\ 0 & 0 & 1 \\ 0 & 0 & 0 \\ 0 & 1 & -1 \end{bmatrix}
$$

7.   <u>Orientation to a Computer</u>

In this report we have been interested in introducing the group ring technique as a way of handling identities. With this in mind, we have limited all calculations in this paper to the group ring on S_3 or S_4. This, we hope, reduced actual computations to a minimum and let the technique be clearly seen. In Table III we gave the representations of S_4 for all 24 permutations. It is possible to look up the matrices and do the computations by hand; the operations involved are not lengthy. In S_5 there are 120 permutations. Each has seven representations. It is extremely time-consuming to do the computations by hand. We store the representations of (12345), (1234), (123), and (12). Every permutation can be expressed as $(12345)^i(1234)^j(123)^k(12)^l$, and we compute them as needed. The computer prints out the representations of

$$\sum_{\pi \varepsilon S_5} \gamma_\pi \pi.$$

8.   <u>Problems</u>

1.   Compute the matrix representation of the identity $(a,a,[c,d]) = 0$ based on the function (R,R,RR). Compare your answer with the representation given in Table V, representation number 4.

2.   Verify the Teichmuller representation in a right alternative algebra which is given in Table VI.

3.   Let R be an associative ring satisfying $x^2=0$ for all $x \varepsilon R$. Using the group ring on S_3, prove that $R^3=0$. Outline a technique to show that if $x^3=0$ for all x in R, then $R^6=0$.

4.   Show that in any (γ,δ) algebra, if $\lambda \neq 0$, $1/2$, or 2, then the additive span of the associators is a two-sided ideal.

5. Let R be an associator dependent algebra defined by Eq. (2). Find necessary and sufficient conditions on the coefficients so that R is associative.

Solution: $(\alpha+\beta+\gamma)+(\delta+\epsilon+\eta) \neq 0$

$(\alpha+\beta+\gamma)-(\delta+\epsilon+\eta) \neq 0$

$(\alpha^2+\beta^2+\gamma^2-\alpha\beta-\alpha\gamma-\beta\gamma)-(\delta^2+\epsilon^2+\eta^2-\delta\epsilon-\delta\eta-\epsilon\eta) \neq 0$

6. In a nonassociative ring R, an alternator is an associator of the form (x,x,y), (x,y,x) or (y,x,x). Let A be the additive span of all alternators. Deduce what additional hypotheses are needed in a right alternative ring so that A will be a left ideal. Solution: The alternators form a left ideal if and only if $(a,b,ab) \in A$ for all $a,b,\in R$.

Acknowledgement

— This paper was written while the author held a grant from the Iowa State University Science and Humanities Research Institute.

References

[1] D. Floyd and I. Hentzel, *An application of group rings to associator dependent algebras,* unpublished.

[2] I. Hentzel, *The characterization of (-1,1) rings,* J. Algebra 30(1974), 236-258.

[3] I. Hentzel, G. M. Piacentini, and D. Floyd, *Alternator and associator ideal algebras,* Trans. Amer. Math. Soc.,

[4] I. Hentzel and G. M. Piacentini, *Simple (γ,δ) algebras are associative,* J. of Algebra, to appear.

[5] K. Hickey, *The representations of S_5 and S_6,* unpublished.

[6] C. Maneri, *Simple (-1,1) rings with an idempotent,* Proc. Amer. Math. Soc. 14(1963), 110-117.

[7]   E. Kleinfeld, F. Kosier, J. M. Osborn, and D. Rodabaugh,
      The structure of associator dependent rings,   Trans. Amer.
      Math. Soc. 110(1964), 473-483.

Department of Mathematics
Iowa State University
Ames, Iowa   50010

AN UNNATURAL ATTACK ON THE STRUCTURE PROBLEM
FOR THE FREE JORDAN RING ON 3 LETTERS:
AN APPLICATION OF QUAD ARITHMETIC

B. T. Smith and L. T. Wos

Abstract

We introduce herein a method which yields answers to certain
questions concerning the free Jordan ring, denoted by R, on 3
letters -- questions which were, until now, open. Among others,
we answer the question of dimensionality for each of a number of
the interesting subspaces of R. We also give the dimension of the
kernel for these subspaces, and explicitly present the kernel element for the subspace of degree type $(3,3,2)$.

The method by which this information is obtained departs
sharply from the natural approach which is based on the consideration of (in many cases very large) systems of linear equations.
Rather, our approach rests on two algorithms which avoid much of
the arithmetic difficulty inherent in processing linear equations.
Of greater significance, however, is the fact that, even with the
aid of a high-speed digital computer, the large systems of equations encountered in studying certain subspaces require so much
time and memory that the problem becomes intractable. The subspace of type $(3,3,3)$, for example, requires for the natural
treatment the processing of a system of 5118 equations in 2391
variables -- a truly awesome matrix.

Of special interest is a new arithmetic employed by one of
the two algorithms which, together with the underlying data structure, makes the problem tractable. This arithmetic, called "quad
arithmetic", and the identity on which it is based are first described in this paper. Were it not for quad arithmetic, the cost

of coping with the redundancy encountered in the study of many of
the interesting subspaces of R would be prohibitive.

The remaining topic of major importance is the aforementioned
data structure. The data structure is an indispensable feature
of the computer program from which our results are obtained.
Through its use the program totally avoids the lengthy and cumber-
some symbolic manipulation inherent in a straightforward treatment
of the free Jordan ring on 3 letters. Moreover, a data structure
similar to that described in this paper may well prove useful to
those contemplating an attack on problems in other nonassociative
algebras, especially should that attack entail the use of the
digital computer.

1.   Introduction

Determination of the structure of the free Jordan ring on 3
letters has defied mathematical analysis. The elements of this
ring are finite sums of finite strings of letters which, under
multiplication, satisfy commutativity but not associativity.
These finite letter strings, together with their grouping symbols, are called monomials and are essential to what follows. The
elements of the ring satisfy the linearized Jordan identity. This
identity does not apply to monomials of degree less than 4, where
the degree of a monomial is the number of occurrences of letters
therein.

In this paper we present the results of a study of this free
Jordan ring -- a study conducted with the aid of a high-speed
digital computer. On the one hand, therefore, certain sections
such as Sections 4 to 6 of the paper will be primarily of interest
to the mathematician. Here will be found answers to certain previously unanswered questions concerning the existence and independence of kernel elements, for example. It is here that a new
identity, relating instances of the linearized Jordan identity,

is introduced and proven — an identity which has a profound effect on the ease with which the results are obtained.

On the other hand, Sections 2 and 3 will be of primary interest to the computer scientist. In Section 2 the discussion centers on the data structure which treats monomials (finite strings of letters) as integers thereby totally avoiding the corresponding symbolic manipulation. Section 3 contains the algorithms which enable the program to circumvent much of the difficulty inherent in dealing with large systems of linear equations. One such system, for example, encountered in this study corresponds to a 5118×2391 matrix, where the matrix is far from being sparse. As a problem in linear equations, this system is virtually beyond consideration because of the time and/or space requirements. With the algorithms of Section 3, however, this system yields its information in five minutes of time and two million bytes of memory on an IBM 370/195.

Sections 2.3 and 2.4 are of general interest. In Section 2.3, certain canonicalizations and a natural approach to investigating the dimension of subspaces of the free Jordan ring on 3 letters are described. In Section 2.4, an overview of the approach upon which the work described in the present paper rests is given.

The organization of this paper is chosen with two objectives in mind. First, in order to permit one to skip over certain sections of the paper, an attempt is made to make portions of the paper self-contained. One who is interested in the formal mathematics underlying this effort can therefore, with the exceptions of Sections 2.3 and 2.4, omit reading Sections 2 and 3. On the other hand, if one is chiefly interested in the use of the digital computer in solving problems in abstract algebra and furthermore solving them in a surprisingly efficient manner, Sections 2 and 3 will be of primary concern.

The second objective of the paper is that of describing our attack on the structure problem with sufficient detail that one can apply a similar approach to problems in other algebras, especially those which are nonassociative. With this in mind, we begin in Section 2 with a brief review of the problem (set of subproblems) to be solved. In each case, the object is to find the dimension of a subspace generated by all monomials of a given degree type (i,j,k), where the degree type (i,j,k) of a monomial m gives respectively the number of occurrences of a, of b, and of c in m. In order to understand the attendant difficulties, we turn to a discussion of the natural approach to solving the subproblems. Then, to begin the presentation of the method which avoids these difficulties, we start with the data structure. The data structure is a realization of a function, called a closing function, which maps monomials to positive integers and thus permits symbolic manipulation to be replaced by operations on the integers. Closing functions are informally but fully described.

In the next section, Section 3, we give two algorithms which enable the program to extract the information contained in the systems of linear equations -- the systems arising from using the natural approach -- but without dealing directly with those systems. Since the systems which yield answers to previously open questions are very large, it is essential that their use be circumvented. The first algorithm provides the mechanism for generating appropriate equations, to be considered singly, from which the relevant data is extracted. The second algorithm is even more important in that through its use so-called redundant equations are discarded without ever being generated. Since the redundant equations, those which are linear combinations of ones already examined, occur approximately two-thirds of the time in the systems corresponding to interesting subproblems, the removal of such equations contributes markedly to solvability. The second algorithm is based on "quad arithmetic" which in turn owes its existence to the identity denoted by E_0.

Section 4 contains the formal treatment of closing functions and their properties. Here also one finds the proofs of the lemmas

of Section 2 which give the crucial characteristics of these
functions.

Quad arithmetic and the identity E_0 are the focus of attention in Section 5. A quad is an ordered set of 4 positive integers
which are images under a particular choice of a closing function
of 4 chosen monomials. The quads can be partitioned into classes
such that the members of a class generate a finite-dimensional
vector space over the rationals. The defining characteristic for
each class of quads is the analogue of the concept of degree type
for monomials. Just as the linearized Jordan identity establishes
various relations among the monomials in the class of those of
some given degree type, so also do various identities establish
relations among the quads. The set of identities, the most powerful of which is denoted by E_0, determines the dimension of each
of the vector spaces of quads. Since the quads of a given class
are in one-to-one correspondence with the equations for the same
degree type obtained by substitution into the linearized Jordan
identity, and since such a set of equations can contain an inor-
dinate number of redundant (as defined earlier) equations, we
prefer to work when possible with the quads rather than with the
equations themselves. The reasons for this preference are first,
an impressive gain in efficiency, and second, the total avoidance
of arithmetic problems. The mechanism of computing with the
quads rather than with the corresponding equations (instances of
the linearized Jordan identity) is "quad arithmetic". Because of
this arithmetic certain quads can be directly added or directly
multiplied by a scalar to yield other quads. For each vector
space of quads, employment of the arithmetic yields a set of quads
which contains a basis for the space and contains in addition a
few unneeded quads -- unneeded in that they are linearly dependent. The linear dependence of these unneeded quads is established by a corresponding substitution into the linearized Jordan
identity.

The significance of these quad bases for the structure problem of the free Jordan ring on 3 letters lies in the relation each bears to a corresponding subsystem of linear equations. The complement of each such subsystem consists entirely of redundant equations. Therefore, quad arithmetic plays an essential role in the obtainment of the results presented herein concerning the free Jordan ring, for this arithmetic provides the means for efficiently discarding the redundant equations -- those equations that comprise approximately two-thirds of the set remaining after canonicalization. Section 5 concludes with the proof of the identity E_0 -- the identity upon which quad arithmetic mainly rests.

Finally, the results obtained thus far through the application of the approach described herein are given in Section 6. They include the dimension of each subspace of the free Jordan ring through degree 9. We also give the dimension of the kernel in each such subspace and, for the subspace of degree type (3,3,2), present the kernel element. We then discuss here the choice of basis in certain subspaces. Because in many of the subproblems the basis can be chosen in a consistent manner, we are led to making certain conjectures. We conclude this section with some open questions arising directly from this effort.

2. Problem Specification, Closing Functions, and the Data Structure

The tone of this entire section is rather informal. We begin in Section 2.1 by giving all of the properties and concepts for the free Jordan ring on 3 letters which are employed in the remainder of the paper. We then in Section 2.2 specify the problem, which for us is a set of subproblems, to be solved. There exists a natural approach to solving all such problems, but the approach is, for the interesting cases, virtually unmanageable. To understand the inherent difficulties which therefore led us to the formulation of the method presented here, we discuss in Section 2.3 the natural approach. We discuss there various canonicalizations. This is immediately followed in Section 2.4 by an overview of the approach by which we obtain solutions to the various subproblems -- a method which contrasts sharply with the natural one.

We then turn in Section 2.5 to a description of the data structure upon which our approach rests. Since the properties of the data structure reflect those of certain special functions called "closing functions", we discuss in Section 2.6 quite informally but in detail closing functions and their basic properties. After giving in Section 2.7 examples of the various basic properties, we state without proof in Section 2.8 a number of lemmas and corollaries for closing functions whose proofs appear in Section 4. These results establish for the functions in question the presence of certain additional valuable properties. These additional properties are captured by the data structure and affect the time required to obtain the solution to the various subproblems in a number of ways. We conclude this section with Section 2.9 which discusses the impact on program efficiency of employing a data structure based on closing functions.

2.1 The Free Jordan Ring on 3 Letters

The first concept of importance is that of monomial. A monomial is a finite string of letters, from among a, b and c, and appropriately paired grouping symbols. Some examples are:

a   b   c   ba   ab   cc   (cc)b
((ba)a)a   ((aa)b)a   ((aa)a)b   (ba)(aa) .

The more relevant basic laws are those of commutativity and associativity of addition, commutativity of multiplication, and distributivity. Multiplication is not associative.

The object of study is the free Jordan ring R on 3 letters, a, b, and c. The elements of R are all finite sums of monomials with rational coefficients. The crucial property of R is that its elements satisfy the Jordan identity. The form of this identity that is used throughout the paper is the linearized Jordan identity. The linearized Jordan identity, often denoted by LJI throughout the remainder of this paper, is given by

$$((yz)w)x + ((xz)w)y + ((xy)w)z = (wx)(yz) + (wy)(xz) + (wz)(xy),$$

where w, x, y, and z take on values from the Jordan ring R.

Associated with each monomial are its degree and degree type. The degree $d(m)$ of the monomial m equals the number of symbols in m which are letters. The degree type $dt(m)$ of m is a triple of integers (i,j,k), where i is the number of a's occurring in m, j the number of b's, and k the number of c's. Therefore, if a monomial has degree type (i,j,k) for some i, j, and k, its degree is $i+j+k$. In particular, the last four monomials in the earlier list of monomials all have degree type $(3,1,0)$ and all have degree 4.

We can now define the subsets $S(i,j,k)$ of R which give rise to the set of subproblems described in Section 2.2. For each triple (i,j,k) of integers, $S(i,j,k)$ is the set of all monomials in R of degree type (i,j,k). For each $S(i,j,k)$ there exists a vector space $V(i,j,k)$ over the rationals consisting of all finite sums of elements from $S(i,j,k)$ with rational coefficients. For the study of the $V(i,j,k)$, we find it convenient to have all computation take place within the integers. There is no loss in generality by doing this, for any relation with rational coefficients among monomials can trivially be transformed into a corresponding relation with integral coefficients. Conversely, any relation with integral coefficients expressing a multiple of a given monomial m in terms of other monomials can be converted to a relation with rational coefficients expressing m itself in terms of other monomials by simply dividing by the coefficient of m.

2.2. The Specific Set of Problems to be Solved

Since each of the V(i,j,k) is a vector space, we can ask of
any V(i,j,k): What is its dimension? Were it not for the
linearized Jordan identity, the answer would be quite quickly obtained as follows. Let V(i,j,k) be some given vector space in R.
Then by definition there exists an S(i,j,k) such that the elements
of S(i,j,k) span V(i,j,k), that is, all elements of V(i,j,k) are
expressible as linear combinations of elements from S(i,j,k) with
rational coefficients. Choose any monomial m in S(i,j,k), and
remove from S(i,j,k) all monomials m' such that m' = m under
commutativity of multiplication but where m' is, of course, not m
itself. Since it is trivial to decide for any two distinct monomials whether one can be obtained from the other by repeated
applications of commutativity, and since S(i,j,k) is finite, one
immediately finds a subset U(i,j,k) of S(i,j,k) such that the
elements of U(i,j,k) are linearly independent. We thus would
have, if the LJI were not present, the dimension of V(i,j,k) equal
to the number of elements in U(i,j,k) and as a bonus would also
have a basis for V(i,j,k), namely, U(i,j,k) itself.

For a number of subspaces V(i,j,k), however, the linearized
Jordan identity transforms the problem of determining dimension
from the very straightforward to the very difficult. There are
two reasons for this. First, the linear dependencies established
by the identity are not the trivial type discussed in the previous
paragraph. Second, even with canonicalization, the number of monomials that must be considered in the search for a basis may be
large. In the dimension problem for V(3,3,2), for example, there
remain 3157 monomials after discarding monomials by use of canonicalization for commutativity of multiplication. When this number
is further reduced by canonicalization for distributivity as discussed in Section 2.5, there remain 732 monomials. Almost two-
thirds of these are found to be linearly dependent, the dependency
being directly traceable to the LJI. This example, V(3,3,2),

suggests the need for departing from the natural approach, which
is described in the next section.

Quite specifically, the subproblems to be solved in this
paper are: For each vector space $V(i,j,k)$ as defined above of
the free Jordan ring R on 3 letters such that $i+j+k \leq 9$, find
both its dimension and a basis for it. In addition we are to
find, for certain sets of subspaces, bases which exhibit a uniform structure. Finally, for convenience, we require all basis
elements to be individual monomials rather than non-unary sums.

2.3. Canonicalization and the Natural Approach for Finding Bases

In this section we solve the basis problem for the very
small vector space $V(3,1,0)$ as an illustration of the natural
approach described here. The basis is obtained with the application of a natural and straightforward approach. The approach is
described in sufficient detail to illustrate the role of the LJI
and, more importantly, to provide the framework for the discussion
of the difficulties inherent in finding bases for interesting
subspaces such as $V(3,3,2)$. Utilization of this natural approach
yields a rectangular matrix of simultaneous linear equations from
which a basis is extracted by using standard techniques for
Gaussian elimination.

The natural approach is more than adequate for subspaces of
R through degree 6. The largest such subspace requires, after
various rather obvious but vital canonicalizations described later
in this subsection, the solution of 87 simultaneous linear equations in 96 unknowns. For certain subspaces of degree 7, however,
the method is merely acceptable and requires access to a large
digital computer such as the IBM 370/195. In addition, for reasons of stability, fixed point, as opposed to floating point,
arithmetic must be employed for these large degree 7 cases. Consider, for example, the vector space of degree 7 of type $(3,2,2)$.

Even after the size of the system has been reduced through canonicalization, one is confronted with a 326×248 matrix.

The approach itself, which we apply in the last half of this
subsection to V(3,1,0), can be divided into three steps. First,
by means of substitution into the linearized Jordan identity,
generate the full set of simultaneous linear equations relevant
to the subspace under study. Second, eliminate from the matrix
yielded by the first step those equations which are linearly
dependent on those remaining. Perform the elimination so that
this step not only yields a matrix of full rank but also one
which is upper triangular. Third, where r is the rank of the
matrix of step 2, apply standard Gaussian elimination to yield a
matrix whose first r×r submatrix is diagonal. This last matrix
expresses r dependent monomials solely in terms of the remaining
monomials which, of course, are therefore a basis.

There are certain tacit assumptions present in this three-
step description of the natural approach. To begin with, one
does not apply the approach until certain canonicalizations of
the type alluded to earlier have been applied. The first canonicalization is that with respect to commutativity of multiplication. As was the case in Section 2.2 when passing from $S(i,j,k)$
to $U(i,j,k)$, all monomials but one which are equal under some
number of applications of commutativity are discarded. Since
another of the assumptions is that the approach is applied to
individual monomials rather than to sums of monomials, the unknowns or variables of the equations of step 1 are monomials no
two of which are trivially equal -- equal simply by application
of commutativity of multiplication. An equation from step 1 may,
of course, contain two monomials which are equal, but such non-
trivial equality will only be established by recourse to the LJI.

The second canonicalization is with respect to the linearized
Jordan identity. This form of the identity,

$$((yz)w)x + ((xz)w)y + ((xy)w)z = (wx)(yz) + (wy)(xz) + (wz)(xy)$$

possesses much symmetry. In particular, the identity is completely symmetric in x, y, and z. If not appropriately constrained,
this symmetry will yield in step 1 many duplicate equations. For
example, assume that four monomials are selected such that no two
are identical. Among the possible assignments to be followed by
substitution into the LJI, assume that the first monomial is to
be substituted for w. There are six ways the remaining three
monomials can be assigned to x, y, and z, each of which yields a
linear equation. Because the LJI is symmetric in x, y, and z,
all six equations become identical after canonicalization with
respect to commutativity of multiplication and application of
commutativity of addition. Thus, it is important to constrain the
substitution into the LJI in step 1, and thus be in a position to
interpret "the full set of equations" to mean that the obviously
duplicate equations have been discarded. One canonicalization
rule which avoids these duplicate equations is based on a well-
ordering of the monomials. The rule requires that the substitution be admissible if and only if the monomial being substituted
for x is greater than or equal to that being substituted for y
and that for y is greater than or equal to that for z.

Finally, canonicalization with respect to distributivity is
also assumed. It is not relevant to the example below, but
examples of it are given in Sections 2.5 and 2.9.2. Briefly, this
canonicalization imposes two constraints. First, only those
monomials which are, recursively, products of basis monomials are
admissible for assignment to w, x, y and z in substitution into
the LJI. Second, all variables appearing in linear equations
yielded by step 1 must be, recursively, products of basis monomials. This second constraint is illustrated in Section 2.9.2.
It forces substitution into partial products during the computation of the summands of the LJI.

We are at last ready to apply the natural approach to the
vector space V(3,1,0). The only elements of V(3,1,0) that need
concern us are the individual monomials therein -- the elements

of S(3,1,0) -- because, of course, by finding a basis for
S(3,1,0), we have a basis for V(3,1,0) since V(3,1,0) consists of
finite sums of elements from S(3,1,0). There are twenty elements
in S(3,1,0), sixteen of which are discarded by canonicalization
with respect to commutativity. This leaves us with the set
U(3,1,0), a set of monomials no two of which are provably equal
by simply applying commutativity of multiplication. Let us say
that the monomials which remain are:

$$m_1 = (ba)(aa) \quad m_2 = ((ba)a)a \quad m_3 = ((aa)b)a \quad m_4 = ((aa)a)b.$$

Now to prepare the way for step 1 to generate (without
trivial duplication) the linear equation set, we order the monomials relevant to substitution into the LJI. We choose an ordering such that the monomial b is greater than the monomial a.
This permits canonicalization with respect to x, y, and z in the
LJI by limiting the possible assignments to x, y, and z that we
will now make in step 1 -- limiting the assignments by the requirement that the values for x, y, and z form a non-increasing
sequence.

STEP 1. Generate the appropriate set of equations. First,
find all subsets, each of which consists of four monomials,
that lead to a substitution into the LJI relevant to
U(3,1,0). A set is acceptable if a count of the occurrences
of a in the four monomials yields the number 3, while that of
b's and c's respectively yields 1 and 0. The only admissible
set for type (3,1,0) consists of the monomials b, a, a, and
a. Second, for each admissible set, find all assignments of
the monomials therein to w, x, y and z within the constraint
that the assignment to x is greater than or equal to that for
y is greater than or equal to that for z. There are then the
two assignments: b for w with a for x, y, and z; and, b for
x with a for w, y, and z. Now generate, for each assignment,
the corresponding linear equation coming from the LJI, taking care to canonicalize where necessary. Canonicalization

is necessitated by the requirement that each linear equation must mention only elements from $U(3,1,0)$. For example, the summand $(wy)(xz)$ in the linear equation corresponding to the second assignment evaluates to $(aa)(ba)$ which must be canonicalized immediately to $(ba)(aa)$. For the four monomials m_1, m_2, m_3, m_4 given above, the first assignment leads to the equation

$$-3(m_1) + 3(m_3) = 0,$$

while the second assignment leads to the equation

$$-3(m_1) + 2(m_2) + m_4 = 0.$$

The monomials of $U(3,1,0)$ are assigned variable numbers v_1, v_2, v_3 and v_4, and commutativity of addition is applied where necessary. We then have the matrix which step 1 is intended to yield. For this example, let m_p be v_p for $p=1,2,3,4$. Then the resulting matrix of coefficients is

$$\begin{bmatrix} -3 & 0 & 3 & 0 \\ -3 & 2 & 0 & 1 \end{bmatrix}.$$

Note that the substitution into the LJI corresponding to any admissible assignment always yields with appropriate canonicalization a linear equation all of whose unknowns are in the subspace under consideration. This follows from the fact that each summand of the linearized Jordan identity mentions each of w, x, y, and z and each with exponent 1.

STEP 2. Produce an upper triangular matrix of full rank. Application of standard matrix methods to the matrix of step 1 yields

$$\begin{bmatrix} -1 & 0 & 1 & 0 \\ 0 & -2 & 3 & -1 \end{bmatrix}.$$

The matrix obtained in this step expresses the dependency relations, but not necessarily solely in terms of basis elements.

STEP 3. Diagonalize the first r×r submatrix of the matrix of
step 2, where r is its rank. This has already occurred
accidentally in step 2.

The three-step natural approach has thus obtained for
V(3,1,0) the basis consisting of the two individual monomials,
namely, ((aa)b)a and ((aa)a)b. The corresponding dependency relations are:

$$(ba)(aa) = ((aa)b)a$$

and

$$2((ba)a)a = 3(aa)b)a - ((aa)a)b.$$

The equations for the remaining elements of S(3,1,0) are obviously
obtained with commutativity, while those for an element in
V(3,1,0) employ appropriate substitution. Note that in this case
the linear equations from step 1 do contain pairs of distinct but
equal monomials, but that the equality comes from the LJI. Summing up: S(3,1,0) contains twenty elements; the dimension of
V(3,1,0) would be 4, the number of elements in U(3,1,0), were it
not for the LJI; the correct dimension of V(3,1,0) as a vector
space in the free Jordan ring on 3 letters is 2.

Certain remarks concerning application of the natural
approach to the structure problem can now be more easily put.
First of all, the matrices of linear equations become prohibitive-
ly large even with the utilization of the three types of canonicalization, that with respect to commutativity, that with respect
to symmetry in the LJI, and that with respect to distributivity.
From step 1, the matrix for V(2,2,2) is 87×96; for V(3,2,2),
326×248; for V(4,2,2), 1011×568; for V(3,3,2), 1266×732; and for
V(3,3,3), 5118×2391. Even with access to a large computer such
as the IBM 370/195, some of these matrices require much more memory
than is available on a computer of this type.

Secondly, there are the questions of time and arithmetic stability in step 2. Such large systems of linear equations require
much time to yield a matrix of full rank which is also upper

triangular. For arithmetic stability, fixed point, as opposed to floating point, arithmetic would be of great assistance but would most likely be overwhelmed in the interesting subspaces in that large integers will still occur.

Some of the inherent difficulties in the second step come from the surprisingly numerous occurrences of redundant equations. Recall that a redundant equation is one which is linearly dependent on the others. This redundancy is clearly not caused by the symmetry in the LJI, for that has been removed by canonicalization. Among the equations for the space $V(2,2,2)$, there are 39 which are linearly dependent on the remaining and so are redundant by definition. In $V(3,2,2)$ there are 186 such; in $V(4,2,2)$, 659; in $V(3,3,2)$, 815; and in $V(3,3,3)$, 3573. The natural approach confronts this redundancy directly, which is a mistake.

The foregoing demonstrates in part the reason the structure problem has remained so unyielding. It also clearly delineates the difficulties which the "inverted" approach of the next section circumvents.

2.4.   An Overview of the Unnatural or Inverted Approach

In this section we give a brief account of the key points in our approach to the problem of finding bases for subspaces of the free Jordan ring on 3 letters. The approach is designed to totally avoid certain of the inherent difficulties in the natural approach and greatly reduce the others. We shall refer to this new approach either as "inverted" or "unnatural". One reason for doing so is that the "inverted" approach starts with a set of monomials, chooses a member of that set for analysis, and then seeks an equation based on that analysis which relates the monomial under consideration to the remaining monomials. The natural approach, on the other hand, starts with an equation set, performs various matrix operations on the set, and then finds monomials

with appropriate properties by examining each resultant equation.
The "unnatural" approach relies heavily on an arithmetic, "quad
arithmetic", employing ordered quadruples of integers and in the
main shuns the more familiar matrix methods on which the natural
approach rests. These quadruples or quads correspond to instances of the linearized Jordan identity. Quad arithmetic man-
ipulates these quads rather than the instances of the LJI (for
reasons of efficiency) with the object of finding redundant
equations. Finally, the "unnatural" approach discards symbolic
manipulation in favor of the use of "closing functions", discussed informally in Section 2.6 and formally in Section 4. Thus
the monomials are not treated with their natural representation.

The inverted approach consists of five major steps. As was
the case with the natural approach, here also is the assumption
of canonicalization with respect to commutativity of multiplication, of canonicalization with respect to x, y, z symmetry in the
LJI, and of canonicalization with respect to distributivity. Only
individual monomials rather than sums are considered throughout
the inverted approach. The series of steps is applied to one
chosen vector space $V(i,j,k)$ at a time.

STEP 1. Select an ordered monomial set. The monomial set,
$N(i,j,k)$, selected at this point is intended to be the complement, loosely speaking, in $V(i,j,k)$ of a basis $B(i,j,k)$.
The selection of $N(i,j,k)$ can be left to a subprocedure within the inverted approach -- a subprocedure based on a conjecture concerning the structure possessed by the basis of
various $V(i,j,k)$. On the other hand, prior to calling the
inverted approach, $N(i,j,k)$ can be selected by the mathematician to reflect some different conjecture and then
presented as input to the approach.

STEP 2. Generate the appropriate equation. A monomial m
from $N(i,j,k)$ of step 1 is selected, according to the ordering of step 1, and analyzed to yield four submonomials m_1 to

m_4. The monomials m_1 to m_4 have the property that their product, taken in the right order and with the proper association, is precisely identical to m. Consistent with the statement that symbolic manipulation is discarded, the analysis is not of m but rather of an integer denoted by $g(m)$. Also, four integers $g(m_1)$ to $g(m_4)$ are found rather than four monomials m_1 to m_4. From $g(m_1)$ to $g(m_4)$ one forms two ordered sets, called quads, each of which consists of four integers. The quads, denoted by q and q', are each a reordering of $g(m_1)$ to $g(m_4)$. The choice of the reordering depends directly on how the m_p are multiplied and associated to yield m identically, but without having any recourse to the symbolic expressions of m and of m_p. Briefly, q and q' correspond to assignments for w, x, y, z in the LJI. (The details are found in Section 3.2.) The object of this step, the equation appropriate to the choice of m, is generated by substitution into the LJI. The equation is not a single substitution instance of the LJI -- does not correspond to a substitution for w, x, y, z -- but rather is the difference of two substitution instances and corresponds to q-q'. These differences which correspond to q-q' for some q and q' are preferred to single substitution instances of the LJI because they reflect a conjecture concerning the structure of certain classes of the $V(i,j,k)$.

STEP 3. Discard redundant equations but without recourse to equation generation. This step depends completely on quad arithmetic. With a restricted form of that arithmetic, linear equations are solved, where possible, one at a time. The coefficients in each equation are integers, and the variables or unknowns are quads. This arithmetic is the mechanism for adding and/or multiplying by integers various sets of quads. Each time one of these linear equations is successfully solved, a quad, and hence a substitution instance of the LJI, is discarded as redundant -- linearly

dependent on quads already processed. The redundant instances, quads, discarded by this step require no substitution into the LJI. For example, as an instance of the
identity E_0 we have the quad $(6,1,1,1)$ is equal to 3 times
the difference of the quads $(4,2,2,1)$ and $(1,4,2,2)$. Using
the table at the beginning of Section 2.5, this translates
to the statement that the equation obtained from the
respective substitution into the LJI of bb, a, a, a for w,
x, y, z is equal to 3 times that equation obtained from the
substitution of aa, b, b, a for w, x, y, z minus 3 times
that equation obtained from the substitution of a, aa, b, b.

STEP 4. Generate the remaining equations to test for full
rank. The concern at this point is for those quads which
have not yet been accounted for by steps 2 and 3. In other
words, there may still exist substitution instances of the
LJI which have neither been used to give a relation for a
monomial from $N(i,j,k)$ nor shown to be linearly dependent on
quads already processed. In this step all quads, relevant
to the $V(i,j,k)$ under study, which have not yet appeared in
steps 2 or 3 are generated. Each such quad is considered in
turn by deriving the corresponding substitution instance of
the LJI. It is here that we allow for the possibility that
$N(i,j,k)$ may not be the complement of a basis. At the conclusion of this step we have in effect the upper triangular
matrix of full rank similar to that of step 2 in the natural
approach.

STEP 5. Diagonalize the matrix of step 4. The full rank
matrix, yielded by step 4, because of the properties exhibited in Section 3 of the inverted approach, contains a large
submatrix which is already diagonalized. This step is
accomplished essentially by standard Gaussian elimination on
the set of relations, coming from step 4, among the monomials. The result is the same as that from step 3 of the

natural approach, namely, a set of relations expressing
dependent monomials solely in terms of basis monomials. In
the case where N(i,j,k) is actually the complement of a basis,
there will be a one-to-one correspondence between the relations yielded here and the elements of N(i,j,k). Each element
of N(i,j,k) will then appear in exactly one relation.

The "unnatural" approach employs a data structure, to be described in the next section, which is a realization of a "closing
function". In Section 2.8, closing functions are shown to have
certain valuable properties. These properties are utilized in
steps 2 to 5 of the unnatural approach to further reduce the time
and space requirements for the study of certain V(i,j,k). The key
property is that of "closure" with respect to particular subgroups
of S_3, the group of all permutations on three letters; see
Sections 2.6, 2.8 and 2.9.

2.5. The Data Structure: Avoidance of Symbolic Manipulation

We now come to the data structure whose primary function is
the representation of individual monomials. The nature of the
representation perhaps best justifies terming our approach
"unnatural". The choice of data structure is motivated by the
wish to totally avoid symbolic manipulation and its problems. The
monomials are, therefore, represented by integers rather than by
their natural symbolic expressions -- sequences of letters and
appropriately paired grouping symbols.

We begin immediately with an example, taken from the data
structure, in which the first 27 monomials of interest are represented (Table 1). Examination of the example suggests certain
questions concerning: the well-ordering of monomials, canonicalization, domain and range, specific uses of such a data structure,
additional properties as yet not illustrated, and finally the
abstract mathematical entity of which it is an example. In this

Table 1

m	g(m)	d(m)	dt(m)	g(m_f)	g(m_s)
a	1	1	1,0,0	-	-
b	2	1	0,1,0	-	-
c	3	1	0,0,1	-	-
aa	4	2	2,0,0	1	1
ba	5	2	1,1,0	2	1
bb	6	2	0,2,0	2	2
ca	7	2	1,0,1	3	1
cb	8	2	0,1,1	3	2
cc	9	2	0,0,2	3	3
(aa)a	10	3	3,0,0	4	1
(ba)a	11	3	2,1,0	5	1
(aa)b	12	3	2,1,0	4	2
(ba)b	13	3	1,2,0	5	2
(bb)a	14	3	1,2,0	6	1
(bb)b	15	3	0,3,0	6	2
(ca)a	16	3	2,0,1	7	1
(aa)c	17	3	2,0,1	4	3
(cb)a	18	3	1,1,1	8	1
(ca)b	19	3	1,1,1	7	2
(ba)c	20	3	1,1,1	5	3
(cb)b	21	3	0,2,1	8	2
(bb)c	22	3	0,2,1	6	3
(ca)c	23	3	1,0,2	7	3
(cc)a	24	3	1,0,2	9	1
(cb)c	25	3	0,1,2	8	3
(cc)b	26	3	0,1,2	9	2
(cc)c	27	3	0,0,3	9	3

section we shall answer these questions but not precisely in the
order they occur.

The data structure consists of cells, one for each monomial
represented therein. Each cell contains explicitly the degree
$d(m)$ of the monomial m being represented, the degree type $dt(m)$ of
m, an integral equivalent of the first factor m_f of m, an integral
equivalent of the second factor m_s of m, and a triple whose components reflect the status (basis or non-basis) of m. In addition,
each cell contains implicitly the monomial m itself and the
representation of m in the form of an integer greater than 0.
Thus the example of the first 27 monomials of interest is not
completely faithful to the data structure in that no explicit mention of the monomials or their integral representation is ever
present in the cells. The example also fails to illustrate the
triple reflecting monomial status.

These cells form an array in which the index of a given cell
is the integral value assigned to or representing the corresponding monomial m. These indices are the values $g(m)$ for some
closing function g. The monomial m itself can only be obtained by
recursive use of m_f and m_s and by use of the terminal conditions
that 1 represents a, 2 represents b, and 3 represents c. Here a,
b, and c are the monomials and not the single letters. The
implicit information contained in the data structure is explicitly
given in the 27-monomial example to provide transparent illustrations for use in the following detailed discussion of the data
structure.

We begin by representing the monomials a, b, and c respectively by 1, 2, and 3. This beginning appears innocent enough but in
fact already partially determines for reasons of consistency the
choice of an ordering of the monomials. Recalling that the degree
$d(m'')$ of the monomial m'' is the number of occurrences of letters
in m'', the first property of the ordering is that which requires
m to be less than m' when $d(m)$ is less than $d(m')$. For convenience
we rewrite this property as, $d(m) < d(m')$ implies $g(m) < g(m')$.

Here $g(m'')$, for a monomial m'', obviously denotes the positive integer representing m'', and we say g maps m'' to $g(m'')$.

The second property of the ordering extends the ordering from monomials of different degrees to those of the same degree but different degree types. The degree type of a monomial is a triple of integers giving respectively the number of occurrences of a, b, and c therein. Let m and m' be such that $d(m) = d(m')$. Let $dt(m) = (i,j,k)$, and $dt(m') = (i',j',k')$. If $k < k'$, then $g(m) < g(m')$, and the matter is settled. If $k = k'$ and $j < j'$, then $g(m) < g(m')$. This property may seem needlessly complicated but is in fact a natural extension of mapping respectively a, b, and c to 1, 2, and 3. It should also be noted that the obvious generalization of this property to cover monomials with $d(m) \neq d(m')$ does not achieve the objective of property 1. Such a move would, for example, have monomials of type $(5,0,0)$ placed earlier in the ordering than those of type $(2,1,1)$. The preferred ordering more easily permits interruption and resumption of the study of the vector spaces $V(i,j,k)$ at arbitrarily selected points.

As expected, the third property of the ordering extends it to differentiate between monomials of the same degree type -- monomials of degree type (i,j,k) with $i \geq j \geq k$. But here the rule is rather far from straightforward, for it does not apply to all types (i,j,k). The important effects of restricting its scope do not emerge until Section 2.9, which discusses efficiencies of the data structure. Assume that m and m' are both of type (i,j,k) with $i \geq j \geq k$. Let m_f and m_s denote respectively the first and second factors of m, and m_f' and m_s' denote respectively those of m'. Then $m = m_f m_s$ and $m' = m_f' m_s'$. Since multiplication is commutative, we make one further assumption to have the ordering well-defined. Assume that m_f and m_f' are such that $g(m_f) \geq g(m_s)$ and $g(m_f') \geq g(m_s')$. When $g(m_s) < g(m_s')$, we require $g(m) < g(m')$. When $g(m_s) = g(m_s')$ but $g(m_f) < g(m_f')$, we also require $g(m) < g(m')$.

Examination of the 27-monomial example shows that the extension to all (i,j,k) of this third ordering property is already violated in type (1,2,0). The monomials (ba)b and (bb)a are respectively mapped under g to 13 and 14 which is contrary to an extension of property 3. Since their second factors are respectively mapped to 2 and 1, it would have been necessary to map (ba)b and (bb)a respectively to 14 and 13. By choosing to retain property 3 as given, (that is, not having property 3 extend to all (i,j,k)), we exchange total uniformity of the ordering for that preserving "permutation image". For example, if one takes the monomials 11 and 12 of type (2,1,0), applies the permutation ab to each, and then applies commutativity of multiplication, one obtains the monomials of type (1,2,0) and in the desired order and respectively mapped to 13 and 14 as is the case. Summarizing, the order of the monomials whose degree type (i',j',k') fails to satisfy i' $\geq$ j' $\geq$ k' is completely determined by the monomials whose degree type (i,j,k) does satisfy i $\geq$ j $\geq$ k, where (i',j',k') is a permutation of (i,j,k).

This last application of commutativity for "permutation image" and the proviso in the definition of property 3 that first factors be represented by larger ($\geq$) integers than second factors are both examples of canonicalization with respect to commutativity of multiplication. This observation leads naturally to the general question of canonicalization which in turn brings to mind the corresponding discussion in Sections 2.2 and 2.3. There U(i,j,k) was selected to canonicalize for commutativity of multiplication also. As one may have already suspected from a glance at the sets of type (1,1,0) and (2,1,0) in the 27-monomial example, canonicalization is immediately realized with the data structure, at least as far as commutativity of multiplication is concerned. And, of course, canonicalization of distributivity is realized just as simply by use of the data structure.

To illustrate the data structure's role in automating distributivity, we present ten additional cells.

m	g(m)	d(m)	dt(m)	g(m$_f$)	g(m$_s$)	Monomial coef.	Status ptr.	length-1
((ba)a)a	30	4	3,1,0	11	1	2	3	1
((aa)b)a	31	4	3,1,0	12	1	0	0	0
((aa)a)b	32	4	3,1,0	10	2	0	0	0
(ba)(aa)	33	4	3,1,0	5	4	1	2	0
(((aa)b)a)a	105	5	4,1,0	31	1	0	0	0
(((aa)a)b)a	106	5	4,1,0	32	1	0	0	0
(((aa)a)a)b	107	5	4,1,0	28	2	0	0	0
((ba)a)(aa)	108	5	4,1,0	11	4	1	24	0
((aa)b)(aa)	109	5	4,1,0	12	4	1	25	2
((aa)a)(ba)	110	5	4,1,0	10	5	1	28	0

The first point to be made is that these monomials are the
full sets of types (3,1,0) and (4,1,0) that are to be represented
as integers. Contained in this remark is the implication that
additional canonicalization is taking place, that for distributivity. To aid understanding here, we make the second point,
namely, bases have already been chosen in both V(3,1,0) and
V(4,1,0). The choice in V(3,1,0) immediately restricts the possible choices in V(4,1,0), when the unnatural approach is employed,
because of the manner in which the data structure handles
distributivity.

The rule for canonicalization of distributivity is: a monomial m = m$_f$m$_s$ is mapped to a positive integer if and only if both
m$_f$ and m$_s$ are basis monomials in their respective subspaces.

This rule is implemented in the data structure with the con-
vention that, a zero occurs in the column labeled coefficient if
and only if the corresponding monomial is a basis element. Thus,
in this example, ((ba)a)a and (ba)(aa) are not basis monomials
while ((aa)b)a and ((aa)a)b are. Then we have, with the canonicalization rule for distributivity, the monomials (((ba)a)a)a and
((ba)(aa))a are not represented in type (4,1,0) and so, of course,
are not eligible for consideration in the search for the basis of
V(4,1,0). On the other hand, the monomials (((aa)b)a)a and

(((aa)a)b)a must be mapped and are, therefore, eligible. In general, the representative for a monomial is chosen by selecting the
next available integer.

We can now answer the question of domain and range for g.
The domain is a proper subset of the monomials in the free Jordan
ring on 3 letters and is equal to the union of subsets denoted by
$R(i,j,k)$. The $R(i,j,k)$ are defined iteratively by the following.
Let $R(1,0,0)$ consist of the single monomial a, $R(0,1,0)$ consist
of b, and $R(0,0,1)$ consist of c. Let g respectively map a, b,
and c to 1, 2, and 3. Then $R(i,j,k)$ with $i+j+k > 1$ is the set of
monomials $m = m_f m_s$ such that

1) both m_f and m_s are basis monomials in their respective
   subspaces, and

2) $g(m_f) \geq g(m_s)$.

The range of g is clearly the positive integers. If the
study being undertaken is that of all $V(i,j,k)$ with $1 \leq i+j+k \leq p$
for some positive integer p, then the $R(i,j,k)$ are generated and
considered in the order dictated by the three ordering properties.
Bases are chosen for $i+j+k \geq 4$ by employment of the unnatural
(inverted) approach. Within the restrictions commensurate with
both canonicalization for commutativity and for distributivity,
the well-ordered subset of monomials is mapped into the positive
integers by choosing the earliest available integer. The mapping
is one-to-one.

The domain and range of g comprise the implicit information
contained in the data structure and discussed in the definition
thereof. There is, however, explicit information therein which
enables one, by iteration, to correctly decide which monomials
will be in the domain of g. Recall that a 0 in the coefficient
column states that the corresponding monomial is a basis element.
The domain of g, therefore, is obtained by taking the union of S_1
and S_2, where S_1 consists of all monomials $m = m'm''$ with m' and
m'' having a zero in the coefficient field and S_2 consists, of
course, of the monomials a, b, and c.

On the other hand, if the coefficient field of some monomial m contains a nonzero integer u, we know that m has a coefficient of u in the relation which expresses m solely in terms of basis monomials. If such is the case, and if $dt(m) = (i,j,k)$ with $i \geq j \geq k$, the pointer field contains the relative address of the equation for m, where the equations are retained in an array, and the length-1 field gives 1 less than the number of basis monomials with nonzero coefficients in the equation for m.

The pointer field has a quite different interpretation for monomials with degree type (i',j',k') which fails to satisfy $i' \geq j' \geq k'$. If m' is such a monomial, its pointer field will contain the value $-g(m)$ for that monomial m with, after canonicalization with respect to commutativity, $m' = t(m)$, where t is the appropriate permutation on 3 letters and $dt(m) = (i,j,k)$ with $i \geq j \geq k$.

For example, $R(3,1,2)$ contains the monomial $m'=((cc)a)((aa)b)$. If the permutation bc is applied to m', one obtains $m'' = ((bb)a)((aa)c)$, which is not admissible for the domain g since $g(m''_f)$ is strictly less than $g(m''_s)$. However, there does exist a monomial m in the domain of g such that m and m'' are provably equal by simply applying commutativity of multiplication. The monomial m is, of course, $((aa)c)((bb)a)$ and is the desired m with $t(m) = m'$ (after canonicalization with respect to commutativity). The monomial m is mapped under g to 543, and so the pointer field of m' contains -543.

Therefore, the data structure through the use of the pointers permits a most valuable move, namely, equations are retained for dependent monomials m if and only if $dt(m) = (i,j,k)$ with $i \geq j \geq k$. As will be discussed in Section 2.9.6, the equation expressing any dependent monomial m' in terms of basis monomials is readily and inexpensively obtained even when $dt(m')$ fails to satisfy $i' \geq j' \geq k'$. The mechanism which makes the obtainment so inexpensive is based on the permutation image property — that property which was chosen in preference to the extension of the third ordering property to cover all triples (i,j,k).

Before turning to the last of the questions posed for this section, we discuss certain conventions which stem from the foregoing material and which are useful throughout the remainder of the paper. First, there is a very strong emphasis and reliance on the sets $R(i,j,k)$ with $i \geq j \geq k$. These will be called "canonical" sets (or "canonical" subsets), and their degree types will be termed "canonical". Second, when referring to some monomial m, there will usually be present the tacit assumption that m is a member of some $R(i,j,k)$, not necessarily canonical. In other words, we are almost exclusively concerned with monomials which obey the rules of canonicalization for both commutativity of multiplication and of distributivity. Thus, if we say that $m' = t(m)$ for some permutation t, we assume that both m' and m are recursively products of basis monomials and that $g(m'_f) \geq g(m'_s)$ and $g(m_f) \geq g(m_s)$. The only exceptions are the monomials a, b, and c and they present no problems. Third, in the discussion of pointers for noncanonical $R(i,j,k)$, an ambiguity exists for the choice of t. If, for example, m' is such that $dt(m') = (1,2,1)$, there are two possible choices of t, the permutations ab and abc. When such ambiguity occurs, we always choose the transposition. An ambiguity can only occur between a transposition and a 3-cycle and only when at least two of i', j', k' are equal. Fourth, and last, we often identify a monomial m with its representation $g(m)$ and will, for example, refer to the monomial 30 to mean the monomial $((ba)a)a$. Here also there is the possibility of ambiguity, for the choice of bases, coupled with the properties of closing functions, completely determines the domain and range of g. More precisely, two choices of bases correspond to two functions g. This remark will be amplified in the next section, Section 2.6. At any rate, we will avoid ambiguity by writing, for example, $R(i,j,k)(g_1)$ where necessary.

Finally, we come to the last of the questions for this section -- of what mathematical entity is the data structure a

realization? The answer is that the data structure is the realization of some "closing function" g, which will be informally
discussed in Section 2.6. The entries therein will vary depending
on the choice of g. However, the number of cells dedicated to the
type (i,j,k), for some given (i,j,k), is a constant -- is independent of the choice of closing function. This number is a function solely of the dimension of each of certain subspaces
occurring earlier in the ordering determined by properties 1 and
2. Put differently, the members of certain R(i,j,k) are a function of the choice of the closing function, but their number is
not.

The R(i,j,k), and hence the data structure, have certain interesting and valuable properties still to come. The discussion
in the next section will provide the background for the emergence
in the form of lemmas and corollaries in Section 2.8. An alter-
nate and formal treatment of the material in Sections 2.6 and 2.8
is found in Section 4.

2.6. Closing Functions: An Informal Presentation

In this section we informally define "closing functions" in
a piecemeal fashion. The formal definition is given in Section 4.
As the properties are encountered, the elementary consequences
thereof are discussed here also.

A closing function is a one-to-one mapping from a subset of
monomials in the free Jordan ring R on 3 letters to the positive
integers. Briefly, closing functions choose a set of monomials,
well-order that set, and canonicalize the members of the set with
respect to both commutativity of multiplication and distributivity. Such a function, denoted by g, by definition maps the
monomials a, b, and c respectively to 1, 2, and 3. In addition
to the exclusion of sums of monomials from the domain of closing
functions, a large fraction of the individual monomials is

excluded from the domain of g. The exclusion of these individual
monomials is the result of two properties.

First of all, if m is a monomial of degree greater than or
equal to 2 such that g is undefined on at least one of the two
factors of m, then m is excluded from the domain of g. Even when
g is defined on both factors m_f and m_s, g will be undefined on
their product m if $g(m_f) < g(m_s)$. As a consequence, one can
easily prove by induction on the degree of the two monomials the
following: If g is defined on m and m', and if m' can be obtained
from m by some number of applications of commutativity of multiplication, then m and m' are precisely identical.

The second property for the exclusion of monomials can be
thought of as designed with the following objective in mind. The
domain of a closing function must not, for example, contain the
monomials m = (((aa)a)a)b and m' = ((((bb)b)b)b)((aa)(aa)). The
point of this exclusion is that m_f and m_s' are linearly dependent
monomials, from a simple application of the LJI, and hence no
basis in the type (4,0,0) exists containing both m_f and m_s'. We
therefore define B(i,j,k)(g), for a given (i,j,k), to be the set
of monomials m in the domain of g such that dt(m) = (i,j,k) and
such that there exists an m' in the domain of g with m and m_f'
identical as monomials or an m" therein with m and m_s'' identical.
For a function g to be a closing function, all B(i,j,k)(g) must be
bases in their respective V(i,j,k). We also require that, if g is
defined on m, g is defined on both m_f and m_s. We further require
of g that, if m and m' are respectively members of B(i,j,k) and
B(i',j',k') with g(m) ≥ g(m'), g be defined on their product mm'.
The latter requirement is needed to guarantee that all R(i,j,k)(g)
are non-empty, where R(i,j,k)(g) is the set of monomials m in the
domain of g with dt(m) = (i,j,k). Each V(i,j,k), therefore, is
represented in the domain of g by a subset containing a basis
B(i,j,k) for that V(i,j,k).

This discussion of the B(i,j,k) brings us back to the remark
in Section 2.5 concerning the possible ambiguity of identifying a

monomial with its representation, its image under g. To discuss
both the question of ambiguity and the determination of the
image, we require a thorough understanding of the additional
axioms, those of ordering, of closing functions. These axioms in
conjunction with those already given will show that: There exist
monomials m, through degree 4, such that g is defined on m for
all closing functions g, and such that $g_1(m) = g_2(m)$ for any two
closing functions g_1 and g_2; there exist monomials m and closing
functions g_1 and g_2 such that m is in the domain of g_1 but not in
the domain of g_2; and, there exist monomials m and closing functions g_1 and g_2 such that m is in the domain of both g_1 and g_2
but $g_1(m) \neq g_2(m)$. The first of these three results is, of
course, a direct consequence of the fact that the LJI does not
apply until degree 4.

The first ordering axiom for closing functions g requires
$g(m) < g(m')$ when $d(m) < d(m')$. Where $dt(m) = (i,j,k)$ and
$dt(m') = (i',j',k')$ with $d(m) = d(m')$, the second requires
$g(m) < g(m')$ when either $k < k'$ or $k = k'$ and $j < j'$. Where
$dt(m) = dt(m') = (i,j,k)$ with $i \geq j \geq k$, the third requires
$g(m) < g(m')$ when either $g(m_s) < g(m'_s)$ or $g(m_s) = g(m'_s)$ with
$g(m_f) < g(m'_f)$.

The fourth and last ordering axiom for closing functions
must be discussed in greater detail. This property is the "permutation image" property of Section 2.5. It states that the
mapping of monomials in noncanonical sets to the positive integers is an induced mapping -- completely determined by the
mapping in the appropriate canonical set. To elaborate, let m
and m' be in R(i,j,k) with $i \geq j \geq k$ and be such that $g(m) <$
$g(m')$. Let t be a permutation, if such exists, on 3 letters such
that the degree type of t(m) is noncanonical. If t_1 and t_2 are
two such with $t_1 \neq t_2$, and if the degree type of $t_1(m)$ equals
that of $t_2(m)$, let t be the transposition in the set consisting
of t_1 and t_2. (One can very quickly prove that, in such a case,

exactly one of t_1 and t_2 is a transposition.) If such a t exists, this fourth axiom first requires the existence of a monomial m'' in the domain of g such that t(m) is obtainable from m'' by simply applying commutativity of multiplication some number of times. Similarly, therefore, there must exist an m''' for m'. The axiom then requires g(m'') < g(m''').

We choose for convenience to refer to such m'' and m''' as t(m) and t(m') for a correctly chosen t and thereby tacitly assume canonicalization. The fourth ordering axiom, therefore, permits the existence of monomials t(m) and t(m') in noncanonical R(i,j,k)(g) such that g(t(m)) < g(t(m')) even with $g(t(m)_s)$ > $g(t(m')_s)$. Such monomials t(m) must, of course, satisfy the two exclusion properties for domains of closing functions. We therefore have $g(t(m)_f) \geq g(t(m)_s)$ and $t(m)_f$ and $t(m)_s$ are members of some B(i,j,k)(g). Thus, if $g(m_1)$ < $g(m_2)$ <...< $g(m_p)$ are the images under some closing function g for $m_1, m_2, \ldots, m_p$ of R(i,j,k) with i $\geq$ j $\geq$ k, then $g(t(m_1))$ < $g(t(m_2))$ <...< $g(t(m_p))$, where $t(m_u)$, for u = 1,2,...,p, are in a noncanonical set and t is a permutation on 3 letters chosen by the rule which prefers transpositions.

Before turning to the consequences of the axioms, we present one last axiom. For a function g to be a closing function, there must exist for any positive integer p a monomial m such that g(m) = p.

We are now ready to discuss the elementary consequences of the ordering axioms. The surprising ones are left to Section 2.8 with the proofs in Section 4. Earlier in this section we mentioned a set of monomials on which all closing functions are defined. That set includes, among others, all monomials listed in the 27-monomial example found at the beginning of Section 2.5. The remaining members are found in the union of R(i,j,k) with i+j+k = 4. In fact all closing functions are identical in their domain and range through degree 4.

Although Section 2.7 contains a number of illustrations directly relevant to a proof of this statement, we shall touch briefly on the essence of the proof here. By definition, all closing functions map a, b, and c to 1, 2, and 3. The monomials of degree 2 must come next, and first among those must come the one of type (2,0,0), namely, aa. The image of aa must be 4 by the last given axiom. Next for consideration, because of the ordering axioms, are those of type (1,1,0). But ba must be mapped and ab must not by the exclusion rules. The permutation image axiom of ordering first has a meaningful effect at type (1,2,0). Skipping ahead, one finds that the choice of closing function g has no effect through degree 4 because the linearized Jordan identity does not apply until degree 4.

At degree 5, however, the choice matters, for there exist closing functions g_1 and g_2 with $B(4,0,0)(g_1) \neq B(4,0,0)(g_2)$, for example. (We should be more pedantic perhaps and say initial segments of closing functions, for there is a certain relevant question of existence to be discussed in Section 6.) Just choose ((aa)a)a in one basis, and (aa)(aa) in the other. Since the dimension of V(4,0,0) is 1, (((aa)a)a)b will be in the domain of g_1 and not in the domain of g_2, while ((aa)(aa))b behaves conversely. The monomials will both be mapped to 103 respectively by g_1 and g_2.

As a different example, there exist closing functions g_1 and g_2 with $g_1(m) = 105$ and $g_2(m) = 106$, where m = (((aa)b)a)a. The former occurs when monomials 31 and 32 of the second table of Section 2.5 are chosen as the basis in R(3,1,0), and the latter occurs when 30 and 31 are chosen as basis.

The choice of the particular closing function g has a marked effect on the magnitude of the coefficients which occur in the equations expressing dependencies in the various R(i,j,k). From the computational point of view, a poor choice of g, in fact, may prevent one from finding a basis for a given R(i,j,k).

2.7. Detailed Illustrations of Closing Functions

The sole purpose of this section is to provide greater familiarity with closing functions. For the properties of Section
2.6, and hence in effect for those of Section 2.5, we employ the
notation $p_0', p_1', \ldots, p_7'$. The formally defined properties
$p_0, p_1, \ldots, p_7$ are given in Section 4.

By p_0' we denote the axiom that closing functions map a, b,
and c to 1, 2, and 3; p_1' to p_3' are the ordering axioms; p_4' is in
effect the requirement that $g(m_f) \geq g(m_s)$; p_5' is the requirement
that $B(i,j,k)(g)$ be basis, that g be defined on each product of
two basis monomials, and that g defined on a monomial implies
that g is defined on its factors; p_6' is the permutation image
axiom; and p_7' is the requirement that $g^{-1}(p)$ exist.

Consider the following lexically ordered set of monomials.

$$m_1 = a \qquad m_2 = b \qquad m_3 = aa \qquad m_4 = ab$$
$$m_5 = ba \qquad m_6 = bb$$
$$m_7 = (aa)a \qquad m_8 = a(aa) \qquad m_9 = (aa)b \qquad m_{10} = (ab)a$$
$$m_{11} = (ba)a \qquad m_{12} = ((aa)a)a \qquad m_{13} = (aa)(aa) \qquad m_{14} = ((aa)a)b$$
$$m_{15} = ((aa)b)a \qquad m_{16} = ((ab)a)a \qquad m_{17} = ((ba)a)a \qquad m_{18} = (ba)(aa).$$

By the numbered steps given here, we show how the domain is
determined and how the monomials are ordered. The monomials m_1
to m_{18} are considered in order.

1) By p_0', $g(m_1) = 1$ and $g(m_2) = 2$.
2) By p_4' and p_5', g is defined on m_3.
3) Therefore, since a and b are basis monomials, p_5' suggests m_4
    for the domain of g. But p_4' excludes it.
4) On the other hand, p_4' permits g to be defined on m_5, and p_5'
    requires it. Commutativity yields m_4 from m_5, so the equi-
    valence class under commutativity of which m_4 is a member is
    represented in g. But p_5' forces many such classes to be
    totally unrepresented in that it requires elements to be re-
    cursively products of basis elements. Therefore, for a

particular choice of g, there will exist equivalences under commutativity which are unrepresented in the domain of g.

5) By p_2', $g(m_3) < g(m_5)$.

6) By p_4' and p_5', g is defined on m_6, and by p_2', $g(m_5) < g(m_6)$.

7) Since it can be shown that $g(m_3) = 4$, p_4' and p_5' will cause g to be defined on m_7, but p_4' excludes m_8. By p_1', $g(m_6) < g(m_7)$.

8) By p_4', p_5', and p_2', g is defined on m_9 with $g(m_7) < g(m_9)$.

9) Since m_{10} has m_4 as its first factor, and since 3) excluded m_4, p_4' excludes m_{10}.

10) By p_4' and p_5', g is defined on m_{11} through m_{15}. By p_3', $g(m_{11}) < g(m_9)$ because the second factor of m_{11} is mapped to an integer smaller than that for the second factor of m_9.

11) While p_1' gives $g(m_9) < g(m_{12})$, it is p_3' which shows that $g(m_{12}) < g(m_{13})$ as in 10). Then p_2' gives $g(m_{13}) < g(m_{14})$ and $g(m_{13}) < g(m_{15})$, but $g(m_{15}) < g(m_{14})$ from p_3' as in 10).

12) Since m_{10} is the first factor of m_{16}, and since 9) excluded m_{10}, g is not defined on m_{16}.

13) We can apply 2), 4), and 10) with p_4' and p_5' to establish that g is defined on m_{17} and m_{18}. By p_3', $g(m_{17}) < g(m_{18})$ and $g(m_{14}) < g(m_{18})$. We use p_2' to show $g(m_{13}) < g(m_{17})$. It is p_3' that gives $g(m_{17}) < g(m_{15})$, but using the argument in 10) on first instead of second factors.

To find the specific integers which are the images under g, it would be necessary to apply in effect p_7' to continually choose the next available integer. The sets must be considered in the order dictated by p_1' and p_2'. Within each set, the monomials must be filtered with p_4' and p_5' and ordered with p_3' and p_6', depending on whether the set is canonical or noncanonical.

2.8. Closure of Bases

In this section we are at last ready for the interesting and

important properties of closing functions which were alluded to
earlier. Without these properties, the results for the interesting subspaces of the free Jordan ring R on 3 letters would not
have been obtained. The key property to be studied is that of
"closed basis".

Before giving the definition, we must introduce additional
concepts. Let g be some given closing function. Recall that
$R(i,j,k)(g)$ is the set of monomials m in R on which g is defined
and for which $dt(m) = (i,j,k)$. Now define $T(i,j,k)$ to be the
set of permutations t on 3 letters such that, for m with
$dt(m) = (i,j,k)$, $dt(t(m)) = (i,j,k)$.

For example, $T(2,1,1)$ and $T(5,2,2)$ are equal and consist of
the transposition bc and the identity; $T(3,2,1)$ consists of the
identity alone; and $T(2,2,2)$ equals S_3, the group of permutations
on 3 letters. All $T(i,j,k)$ are subgroups of S_3.

We then define a basis $B(i,j,k)(g)$ or $R(i,j,k)(g)$ to be
closed when, for all t in $T(i,j,k)$, the monomial m is a member of
$B(i,j,k)(g)$ if and only if (the canonicalization with respect to
commutativity of) $t(m)$ is a member of $B(i,j,k)(g)$. We imply
here, of course, that $B(i,j,k)(g)$ is a subset of $R(i,j,k)(g)$. We
often omit the notation (g) when no confusion arises.

The parenthetical qualifier is present for correctness of
definition, but it shall be dropped from now on. It recalls the
fact that canonicalization of monomials is assumed but does not
connote that the closing function g has an effect on the choice
of canonicalized monomial. In fact one can easily show, because
of the ordering axioms, that only one representative under commutativity of multiplication can ever be mapped regardless of the
closing function under consideration. The $B(i,j,k)$ of the definition turn out to be, of course, the $B(i,j,k)(g)$ of Section 2.6.
There they were used to canonicalize with respect to distributivity. We now return to the main discussion.

The first observation to be made is that, from the very
general viewpoint, not all bases are closed, even for the early

subspaces. In fact, even if sums of monomials are barred from consideration, there exists for V(2,2,0) a basis which is not closed. Let B'(2,2,0) be the basis consisting of

$$m_1 = ((bb)a)a \quad m_2 = ((ba)b)a \quad m_3 = (bb)(aa) \quad m_4 = (ba)(ba).$$

(This basis will prove useful later in this section.) The only permutation to consider is ab. Although ab applied to m_3 and applied to m_4 present no essential problem since canonicalization with respect to commutativity yields m_3 and m_4 themselves, ab of m_1 and ab of m_2 cannot be helped by such canonicalization. In other words, there do not exist in B'(2,2,0) monomials which are commutative variants of either ab of m_1 or ab of m_2.

A more interesting example is that of partial closure. Even with canonicalization for commutativity, there exist examples of individual monomial bases for V(2,2,2) which are closed under ab but not under all of $S_3 = T(2,2,2)$. This brings us to an obvious but useful remark.

Remark 1. If a basis is closed under at least two different transpositions, then it is closed under all of S_3.

We come now to the first major result, Lemma 1, for closing functions. From this lemma we shall see that one has less freedom in constructing a (partial) realization of a closing function than is apparent. Put another way, the sets of ordered pairs, monomials each with its integral image, which correspond to the domain-range of some closing function are fewer than one might expect.

Lemma 1. Let g be any closing function. Then R(i,j,k)(g) is not empty for all (i,j,k) with i+j+k > 0. More importantly, if R(i,j,k), for some (i,j,k), is such that $i \geq j \geq k$, and if B(i,j,k)(g) is that set defined in Section 4 by $p_5(g)$, then B(i,j,k)(g) is closed and a basis for R(i,j,k)(g).

This lemma shows that an attack on the problem of finding bases for the subspaces of R which employs a data structure of

the type described in Section 2.5 is subject to the powerful constraint of closure. Closed bases must be chosen at each step as
one considers the subproblems of the $V(i,j,k)$ in the order given
in Section 2.5. Failure to do so prevents the continuation of
the partially developed structure in a manner consistent with the
desire of realizing some closing function.

To see how the continuation is prevented and glimpse the
disastrous consequences, assume that the non-closed basis
$B'(2,2,0)$ cited earlier in this section is chosen as the basis in
type $(2,2,0)$. By the distributivity rule in Section 2.5, the
monomial $m_1 b$ must be assigned a positive integer. But by the
permutation image property -- by using the pointer field in the
data structure for $m_1 b$ of degree type $(2,3,0)$ -- we would have to
find the monomial equal to ab applied to $m_1 b$. But this is just
$ab(m_1)a$, which implies by the distributivity rule that $ab(m_1)$ is
a basis monomial. In fact, as was pointed out, no commutative
variant of $ab(m_1)$ is in $B'(2,2,0)$. One might conjecture that a
reasonable countermove would be to adjoin the needed missing
monomials to the domain of definition -- have present in the data
structure, for example, $ab(m_1)a$. Unfortunately, by the time the
subproblem of $V(3,3,2)$ comes up, the number of such forced
adjunctions propagated into the $(3,3,2)$ space is roughly the number which are naturally found there as the result of the use of
closing functions.

Resuming the discussion of closing function properties and
their consequences, we have Lemma 2 which shows how well noncanonical sets reflect canonical ones.

Lemma 2. Let g be any closing function, and $R(i,j,k)(g)$ for some
(i,j,k) be such that $i \geq j \geq k$. Let $B(i,j,k)(g)$ be the basis of
$R(i,j,k)(g)$. If t is in S_3 with $t(R(i,j,k)(g))$ noncanonical and
equal to $R(i',j',k')(g)$, then the canonicalization with respect
to commutativity of $t(B(i,j,k)(g))$ is a basis of $R(i',j',k')(g)$
and in fact equals $B(i',j',k')(g)$.

We then have the expected lemma, Lemma 3, concerning T(i,j,k). It states that in effect T(i,j,k) maps R(i,j,k) to itself.

Lemma 3. Let g be any closing function, and consider R(i,j,k)(g) for some (i,j,k). If m is in R(i,j,k)(g) and t is a permutation on 3 letters such that t(m) and m have the same degree type, then the canonicalization with respect to commutativity of t(m) is a member of R(i,j,k)(g).

Now we have the result, Corollary 1, which couples Lemmas 1 and 2 and thereby shows that closing functions, and hence the data structure, behave uniformly for all (i,j,k).

Corollary 1. The bases B(i,j,k)(g) are closed for all closing functions g and all (i,j,k).

Thus one must choose closed bases for the noncanonical sets if one hopes to have a data structure which is in effect a closing function and, therefore, avoids the consideration of many unnecessary monomials.

Finally, we have the closure result for non-bases -- a result which is important in the application of the inverted approach.

Corollary 2. The set of non-basis monomials in R(i,j,k)(g) is closed for all (i,j,k) and all closing functions g.

The lemmas and corollaries of this section establish properties for closing functions whose absence would have prevented the completion of the study upon which this paper is based. In this study we employ the initial segment through degree 9 of a closing function. We have in fact made other studies through degree 9 with a number of distinct closing functions (that is, the corresponding initial segment) and also studies based on non-closed bases which are, of course, equivalent to partial abandonment of closing functions. The employment, for example, of B'(2,2,0) was quite unsatisfactory. The efficiencies resulting from the use of

closing functions are quite startling and occur in a number of
areas, which takes us to the next section.

2.9.   Efficiencies of the Data Structure

Up to this point, we have described many important properties of the data structure. Now in this section, we describe how
these properties lead to efficient processes for solving the
basis and dimension problem for each subspace $V(i,j,k)$ of R. Recall that, although the study is of the free Jordan ring R over
the rationals, all computation proceeds with integral values.
For example, all coefficients of dependent monomials and equations are retained as integers.

2.9.1.   Monomial Representation

Each monomial m is represented in the data structure implicitly as the integer $g(m)$ rather than as a letter string with
grouping symbols where $g(m)$ is the index of the cell in the data
structure characterizing m. This representation has two major
advantages: first, it saves storage space as the cell for each
monomial contains two fixed length integers rather than a
variable length string consisting of n letters and $2(n-2)$ grouping symbols for a monomial of degree n; second, the monomials are
well-ordered by their integer representation thus greatly facili-
tating monomial comparisons used throughout the process of, for
example, canonicalization for commutativity.

2.9.2.   Canonicalization of Monomials

The data structure permits us to work with the smaller subsets $R(i,j,k)$ rather than the full monomial subsets $S(i,j,k)$,

thus achieving a substantial saving in space. For example, the
subset S(5,1,1) contains 5544 monomials whereas the subset
R(5,1,1) in which all the commutative and distributive variants
are removed contains only 55 monomials. By noting that the
corresponding reduction for the degree type (4,1,1) is from 1260
to 34 monomials, the import of dealing with R(i,j,k) instead of
S(i,j,k) becomes evident. Indeed, fewer monomials result in a
marked saving in time since the time required to determine the
dimension of the spaces is approximately proportional to the
square of the number of monomials present.

The next major advantage that the data structure provides is
to efficiently avoid the generation of commutative and distributive variants of monomials. In fact, such variants do not exist
in our representation, and thus there is no price paid to work
with the smaller subsets R(i,j,k). This is accomplished in two
ways and is best illustrated by an example of generating an instance of the LJI in the subset R(4,1,0).

Consider the subset R(4,1,0) and the instance of the LJI
where a is substituted for w, y, and z, and ba for x. From Section 2.5, we see that our data structure assigns 1 and 5
respectively for the monomials a and ba, and in terms of these
integers, the LJI becomes

$$(1,5,1,1): \quad ((1.1).1).5 + ((5.1).1).1 + ((5.1).1).1 =$$
$$(1.5).(1.1) + (1.1).(5.1) + (1.1).(5.1)$$

where . represents product in the ring. The goal is to express
the equation in terms of monomials in R(4,1,0). First, commutative variants are never formed since at each ring product, the
first factor of each product is required to be an integer not
less than the integer representing the second factor. Hence, the
factor 1.5 in the fourth term of (1,5,1,1) is changed to 5.1 before beginning the search for its integer representation in the
table given in Section 2.5. Similarly, the first and second factors of the fifth term (1.1).(5.1) are 4 and 11 respectively, but

their product is determined by searching for 11.4, instead of
4.11, in the second table of Section 2.5. Second, the distributive variants are never formed, since each product representing a
dependent monomial is replaced by its expression in terms of basis
monomials. This is illustrated by the second and third terms
((5.1).1).1 of (1,5,1,1). Here, the first table in Section 2.5
shows 5.1 is 11, and the second table of Section 2.5 shows 11.1
is 30. By consulting the second table of Section 2.5, monomial
30 is seen to be dependent since its coefficient field is nonzero.
The expansion of 30 in terms of basis monomials 31 and 32 in
R(3,1,0) is

$$2(30) = 3(31)-(32),$$

where the integers 2 and 3 preceding monomials 30 and 31 are elements of the ring of coefficients and not images under g of b and
c respectively. Thus, in place of the two monomials ((5.1).1).1
we have the expression

$$3(31.1)-(32.1).$$

Consequently, in this fashion, the distributive variant (30.1) is
never formed.

Summarizing, the avoidance of commutative variants is accomplished simply by integer comparison, and that for distributivity
is accomplished by flag-testing coupled with replacement of single monomials by monomial expressions. The variants are never
formed thus obviating the need for search processes to discard
them.

2.9.3. Canonicalization in the Selection of Instances of the LJI

In a similar fashion to the canonicalization of monomials,
the data structure is instrumental in providing efficient techniques for the canonicalization in the selections of instances of
the LJI. This canonicalization circumvents two kinds of

unnecessary duplication of equations resulting from instances of
the LJI; first, that duplication caused by the symmetry in x, y,
and z of the LJI; and second, that duplication caused by substitution of non-basis monomials for w, x, y, and z in the LJI.

This symmetry in the LJI produces equational duplication in
the following fashion. If given any assignment, say q_1, to w and
any three values q_2, q_3, q_4, the same equation is generated for
all assignments of q_2, q_3, and q_4 to x, y, and z. This can
readily be proven by application of commutativity of multiplication and addition to the terms of the LJI. Since the monomials
are well-ordered by the data structure, the duplicate equations
are easily avoided by insisting that the integer representations
of the assignments to x, y, and z form a non-increasing sequence,
that is, $x \geq y \geq z$. This dispenses with the first kind of
duplication.

The second kind of duplication is not simply that which
occurs in identical equations. Because the LJI is linear in each
of its arguments w, x, y, and z, assignments of non-basis monomials to any of w, x, y, or z will produce equations which are
just linear combinations of those equations obtained from instances of the LJI in which only basis monomials are assigned to
w, x, y, and z. Hence, from the point of view of equation manipulation, the equations generated from non-basis monomial substitution duplicate information about monomial dependency in the
Jordan ring -- information already available from using basis
monomial substitution. Again, using the data structure, we can
readily prevent duplicates from being generated by insisting that
only basis monomials be assigned to w, x, y, and z.

Therefore, for any R(i,j,k), we are only concerned with sets
of four values (q_1, q_2, q_3, q_4), called quads. Furthermore, we are
only interested in quads that are termed canonical, that is,
$q_2 \geq q_3 \geq q_4$ and q_1, q_2, q_3, q_4 are images of basis monomials.

2.9.4. Generation of all Quads, Canonical and Relevant
        to a Given R(i,j,k)

The above two canonicalizations and the data structure now
provide an efficient process for generating the assignments to w,
x, y, and z for instances of the LJI relevant to a specific subset R(i,j,k). An instance of the LJI is relevant to R(i,j,k) if
and only if the coordinatewise sum of the degree types of the
assignments to w, x, y, and z is (i,j,k). The procedure for selecting the quads q = (q_1,q_2,q_3,q_4) which are both canonical and
relevant to R(i,j,k), for a given (i,j,k), is as follows.

Select the largest value in the data structure whose degree
type is coordinatewise less than or equal to (i,j,k), whose
coefficient field is zero, and whose degree is i+j+k-3. A zero
in the coefficient field states that the value corresponds to a
basis monomial. The requirement that the degree equals i+j+k-3
leaves room for the three remaining values to be chosen. Assign
the selected value to q_1. For q_2, choose the largest possible
value such that its degree type plus that of q_1 is coordinatewise
less than or equal to (i,j,k), such that the coefficient field is
zero, and such that the degree of the chosen q_2 plus that of the
chosen q_1 leaves room for the selection of both q_3 and q_4. Thus,
$i+j+k-d(q_1)-d(q_2) \geq 2$. Similarly, select the largest possible q_3
but with the added constraint q_3 be less than or equal to q_2, and
then select the largest possible q_4 but with the constraint that
q_4 is less than or equal to q_3. At this point a quad q has been
found which is both relevant to (i,j,k) and canonical. Additional quads to be retained are found by keeping the values for q_1,
q_2, and q_3, but finding the next smaller value for q_4 other than
that just utilized. The "new" q_4 must, of course, satisfy the
three requirements given above. One finds all possible values
for q_4 in this way which combine with the given q_1, q_2, and q_3.
Then one replaces q_3 by the next smallest appropriate value for
q_3, and resumes the finding of all appropriate values for q_4.

When the available values for q_3 are exhausted, one proceeds in a similar fashion with q_2. Finally, one chooses a "new" and smaller q_1, and applies the procedure just given starting with a "new" q_2.

In the procedure just given, extensive use is made of degree type field, the coefficient field, and the well-ordering of the monomials. In this manner, the data structure provides a very efficient method for selecting the relevant and canonical quads for given $R(i,j,k)$. The set of instances of the LJI which correspond to such a selected set of quads contains all of the information required to determine both the basis of $R(i,j,k)$ and the corresponding dependency relations therein.

At this point, we give a simple example of this algorithm for the subset $R(2,2,0)$. Clearly, the largest value admissible for w is the largest degree 1 variable, whose degree type is coordinatewise less than or equal to $(2,2,0)$, namely 2, which is the image of b. This is because monomials of degree 2 or larger do not leave room for assignments to x, y, and z, and because the assignment of $3 = g(c)$ is clearly inadmissible since the degree type of c is $(0,0,1)$. Since the algorithm requires $q_2 \geq q_3 \geq q_4$, the assignment of 2 to w has only the single completion, that which assigns 2, 1, 1 to x, y, z. In other words, once b is assigned to w, the only assignment to x, y, and z consistent with the demands of being canonical and being relevant is that of b, a, a. The next, and last, choice for w is 1. This choice can also be completed to a canonical quad relevant to $R(2,2,0)$ in only one way, namely, by assigning 2, 1, and 1 to x, y, and z.

Note that as in the canonicalization of the monomials, the duplicate variants of the instances of the LJI are never generated and hence expensive computational sifting processes are not required to filter out the unnecessary equations.

2.9.5. Monomial Searching

Inherent in the generation of the LJI from the specific composite assignment q_1, q_2, q_3, and q_4 to w, x, y, and z is the need to search in the data structure for the integer representations for products, such as $q_4 q_2$. In general, this would be a time consuming process but for the fact that the data structure has certain ordering properties, especially those which order monomials within given degree types (i,j,k). Recall that, if (i,j,k) is canonical, $i \geq j \geq k$, the ordering of the monomials is determined first by the value assigned to the second factor, and second, within equal second factors, by the value assigned to the first factor. On the other hand, if (i,j,k) is noncanonical, the ordering is completely determined by the permutation image property of Section 2.5, which forces the order to depend completely on that of the appropriate canonical set.

The search for the product $q_4 q_2$, where q_4 and q_2 are basis monomials in their respective spaces, proceeds as follows. First, because of the requirement that the first factor be greater than or equal to the second factor, it may be necessary to replace $q_4 q_2$ by $q_2 q_4$. Now assume $q_4 \geq q_2$. Next the degree types of q_2 and q_4 are added coordinatewise to determine the degree type of the product. If the degree type of the product is canonical, then a search process which is faster than linear is used to find $q_4 q_2$. This search takes advantage of the fact that the monomials in canonical sets are ordered first by their second factor and for equal second factors, then by their first factor. We use the Fibonacci search [3] to find the product $q_4 q_2$.

The first and second factors of monomials in noncanonical sets are not ordered in this simple manner which thus requires a more complicated search procedure. Here, we take advantage of the permutation image property that gives the ordering of monomials in noncanonical sets in terms of the corresponding canonical set. Using the degree type field of the product $q_4 q_2$, the correct transformation t in S_3 that maps the corresponding canonical set to the noncanonical set containing $q_4 q_2$ is determined. As noted in Section 2.5, when there is a choice of transformations, the transposition is taken in preference to the three-cycle. Next, using the degree type fields of the data structure and certain transformation tables for basis monomials in canonical sets, the factors q_4 and q_2 are mapped by t^{-1} without the need for a search of the type just described. Then a search as given above is made for the product $t^{-1}(q_4) t^{-1}(q_2)$ in the canonical set. Suppose this product is found to be the p-th monomial in the canonical set. Then the product $q_4 q_2$ is the p-th monomial in the noncanonical set.

Element searches are an integral part of any computational approach to solving the basis problem. For instance, collection of like terms from the substitution into the LJI for w, x, y, and z requires a search procedure to identify identical terms. Our data structure provides a well-ordering to the monomials so that search processes, faster than linear, can be readily implemented.

2.9.6. Generation of Equations

The data structure is instrumental in several ways in providing efficient procedures for generating equations. First, it provides a natural indexing for the monomials. When an instance of the linearized Jordan identity is formed, the coefficients of a particular monomial in an equation can be saved and coefficients of like terms can be collected just by using the g-image of the monomial as an index. This rapid indexing procedure is particularly important as often hundreds of monomials can be generated from one term in the LJI.

Secondly, the data structure is so designed that equations for dependent monomials in a noncanonical set are efficiently derived from the equations for the corresponding dependent

monomials in the canonical set, hence requiring that only the
equations for the dependent monomials in the canonical sets be
retained. To illustrate this procedure, suppose the equation for
a monomial m' in a noncanonical set is desired and m' is the permutation image of m in the corresponding canonical set. (In
fact, the pointer field for m' is set to -g(m) to indicate this
correspondence.) If the equation for m is

$$u \; g(m) = \sum_{p=1}^{n} u_p \; g(b_p)$$

where the b_p are basis monomials in the subspace containing m,
then the equation for m' is just

$$u \; g(m') = \sum_{p=1}^{n} u_p (g(b_p) + d),$$

where $d = g(m')-g(m)$.

In other words, the equations for noncanonical monomials are obtained simply by translating the monomial indices in the equations for the corresponding canonical monomials. This technique
is valid, since for all p, the p-th monomial in the noncanonical
set is the permutation image of the p-th monomial in its corresponding canonical set.

The third way, relevant to equation generation, in which the
data structure provides an efficient technique for solving the
basis problem applies to canonical sets R(i,j,k) for which at
least two of i, j, and k are equal and nonzero. For such sets,
it turns out that the effort in determining a basis can be substantially reduced -- by nearly a factor of two when just two of
the i, j, k are equal and nonzero and nearly a factor of six for
sets where all three of i, j, k are equal. This saving is obtained by taking advantage of two facts: first, whenever e is a
valid equation in, for example, the canonical subset R(i,i,k),
then e' equal to ab(e) is also a valid equation in R(i,i,k) and
is usually independent of e; and secondly, much less effort is
required to derive e' as ab(e) than to derive it from its LJI or

linear combination of instances of the LJI. For example, in some
cases, ab(e) is e itself, after canonicalization.

We consider an example in R(2,2,0) to illustrate the technique. Let $q(m_1,m_2,m_3,m_4)$ be an instance of the LJI in which m_p
for p = 1,2,3,4 are substituted for w, x, y, and z respectively.
The instance q(a,b,b,a) can be rearranged to make dependent the
monomial ((ba)a)b as follows:

q(a,b,b,a): 2((ba)a)b = (bb)(aa) + 2(ba)(ba) - ((bb)a)a.

Applying the transformation ab to the above equation, and canoni-
calizing with respect to commutativity, we obtain the relation
expressing the dependency of the monomial ((ba)b)a as follows:

2((ba)b)a = (bb)(aa) + 2(ba)(ba) - ((aa)b)b.

This same equation, on the other hand, is easily shown to come
from the instance q(b,b,a,a) which is just the canonicalization
of ab of the previous instance q(a,b,b,a).

This discussion of the efficiencies of the data structure
leads naturally to the description of the two main algorithms,
which is the topic of the next section.

3. The Basic Algorithms

In this section we give two basic algorithms which are employed by the inverted approach -- that approach outlined in Section 2.4 and used to obtain the results given in Section 6.1.
For this discussion, we will need certain concepts which are formally treated in Sections 4 and 5, but we will not require such
rigor at this time.

To begin with, for a given (i,j,k), R(i,j,k) is the set of
individual monomials which is the focus of attention for both algorithms. The monomials of this set are all of degree type
(i,j,k), and it is assumed that $i \geq j \geq k$. Each monomial
$m = m_f m_s$ in R(i,j,k) is recursively the product of basis elements

in already chosen bases, which automates the canonicalization of distributivity. Since there do not exist in R(i,j,k) two distinct monomials which can be shown to be equal by application of commutativity of multiplication, and since we are employing the data structure of Section 2.5 with the corresponding properties of closing functions g of Section 2.6, commutativity of multiplication is also automated. For every monomial m under consideration, we thus have $g(m_f) \geq g(m_s)$, where g is the particular closing function currently being employed. Although not all of the preceding sets are needed in the search for a basis in the subspace under consideration, the sets are still considered in the order dictated by properties $p_1(g)$ and $p_2(g)$ in Section 4. For example, as will be illustrated in Algorithm 1, the monomial and basis information for the subspace of type (4,1,0) is irrelevant to the study of R(3,2,1). On the other hand, for our approach we must have the complete information for R(3,1,0), R(2,2,0), R(2,1,1), R(3,2,0), R(3,1,1), and R(2,2,1) to study R(3,2,1). We also need the information for, say, R(1,2,1), but this is obtained from R(2,1,1), as illustrated in Section 2.9, by property $p_6(g)$ of Section 4. We therefore have bases B(i',j',k')(g) for all subsets earlier than R(i,j,k), and we also have the relations expressing the dependent monomials in terms of members from B(i',j',k')(g).

Next we assume that the R(i,j,k) under consideration is partitioned into two sets. The first set, N(i,j,k), will be the complement of a basis if all goes well. The second set R(i,j,k)-N(i,j,k) hopefully is a basis but, even if it is, may be rejected because of certain additional requirements. The most important of these is that R(i,j,k)-N(i,j,k) be closed and also consist of two subsets of monomials, each of which must contain a specified number of elements in order that the efficiencies of Section 2.9 be available. This specification, in terms of "paired" and "fixed" monomials as defined in Section 3.1, reflects the spirit of the last conjecture given in Section 6.2.

Finally, to complete the setting for the presentation of the two algorithms, we briefly discuss the quads of Section 5. A quad is an ordered set of four positive integers which correspond under the chosen function g, to four monomials. The ordering of $q = (q_1, q_2, q_3, q_4)$ gives an assignment to w, x, y, z respectively for a substitution into the linearized Jordan identity. Quads correspond to instances of the LJI. We are only concerned with quads which are both canonical and relevant to $R(i,j,k)(g)$. A quad is canonical if and only if all of q_1 to q_4 are images under g of basis monomials m_1 to m_4 relative to g and $q_2 \geq q_3 \geq q_4$. To be relevant, where $q_p = g(m_p)$, for p = 1,2,3,4, a quad $q = (q_1, q_2, q_3, q_4)$ must be such that the coordinatewise sum of the degree types of m_p equals (i,j,k). In a similar way, we define $q(m_1, m_2, m_3, m_4)$ to be the instance that one obtains by substituting into the LJI m_1 for w, m_2 for x, m_3 for y, and m_4 for z. (In Section 5, we use a similar notation, but at that point, $q(m_1, m_2, m_3, m_4)$ is the expression,

$$((m_3 m_4) m_1) m_2 + ((m_2 m_4) m_1) m_3 + ((m_2 m_3) m_1) m_4 -$$
$$(m_1 m_2)(m_3 m_4) - (m_1 m_3)(m_2 m_4) - (m_1 m_4)(m_2 m_3).)$$

3.1. Selection of N(i,j,k)

As stated in step 1 of Section 2.4, the set N(i,j,k) is an ordered set of monomials in R(i,j,k), selected to be the complement of an expected basis of R(i,j,k). It is not just the set N(i,j,k) that is important here, but the order of the monomials in N(i,j,k) is also crucial. The order is so chosen that the instances of the LJI, selected by Algorithm 1, provide without much computational effort equations which make each monomial in N(i,j,k) in turn dependent upon both monomials occurring later in the ordering of N(i,j,k) and upon monomials expected to be basis elements. The essential property of the equational system generated by Algorithm 1, being used here, is that the matrix representing the equational system is nearly upper triangular.

For those subsets $R(i,j,k)$ such that no two of i, j, and k are equal, the preferred order for the monomials in the corresponding $N(i,j,k)$ is that which is monotonically decreasing with respect to g. In other words, regardless of the particular choice of such an $N(i,j,k)$, the monomials m_p therein are ordered so that $g(m_1) > g(m_2) > ... > g(m_u)$ where g is the closing function currently being used and m_p is the p-th member in $N(i,j,k)$. This ordering is also that which is preferred for the subsets of degree type $(i,0,0)$. On the other hand, when at least two of the i, j, and k are equal and nonzero, the situation is more complex and requires the following digression.

Let $T(i,j,k)$ be the subgroup of transformations in S_3 which maps the degree type (i,j,k) into itself. For example, $T(2,2,0)$ for the subspace of type $(2,2,0)$ consists of the identity and the transposition ab; for the subspace of type $(3,1,1)$, the identity and the transposition bc; for the subspace of type $(2,2,2)$, all of S_3. Now we partition each $R(i,j,k)$ into classes such that each class is closed under all of $T(i,j,k)$. Thus, a given subset C of a given $R(i,j,k)$ is such a closed class of monomials when, for every t in $T(i,j,k)$, the canonicalization of $t(m)$ is in C if and only if m is in C. Each class C is then either termed "paired" or "fixed". A class is termed paired if and only if the number of monomials in C equals the number of transformations in $T(i,j,k)$, and termed fixed if and only if the number of monomials in C is half the number of transformations in $T(i,j,k)$. A monomial m is paired or fixed depending upon the status of the class of which it is a member. We take as the representative of each class of monomials the monomial with the smallest g-image.

As an example of fixed and paired classes, we consider the subset $R(2,1,1)$. In $R(2,1,1)$, there are 9 monomials whose g-images are 50 to 58. The monomials and their g-images are listed below.

g(m)	m	$g(m_f)$	$g(m_s)$
50	((cb)a)a	18	1
51	((ca)b)a	19	1
52	((ba)c)a	20	1
53	((ca)a)b	16	2
54	((aa)c)b	17	2
55	((ba)a)c	11	3
56	((aa)b)c	12	3
57	(cb)(aa)	8	4
58	(ca)(ba)	7	5

Since the T(2,1,1) consists of the identity and bc, each paired class consists of two monomials. There are three such classes, namely, (51,52), (53,55), (54,56). The representatives of these paired classes are respectively 51, 53, and 54. There are three fixed classes, consisting respectively of 50, 57, and 58. In this case, 50, 57, and 58 are the corresponding representatives. On the other hand, the paired classes of R(2,2,2) consist of six monomials each, while the fixed classes consist of three each. There are 11 paired and 10 fixed classes in R(2,2,2).

Now, we are able to specify the preferred ordering of the monomials in N(i,j,k) when at least two of i, j, and k are equal and nonzero. First, N(i,j,k) is required to consist of complete classes C, with paired classes appearing first and fixed classes following. Next, the representative of each class is followed immediately by the remaining members of that class. Finally, based upon the values under g, the representatives are listed in descending order. Thus, for example, if N(2,1,1) consists of (ca)(ba), ((ca)b)a, and ((ba)c)a, the preferred ordering for N(2,1,1) is (51,52,58).

The preferred ordering is not enforced by the computer program, but it is the ordering most often used. In addition, it is the default ordering for the default N(i,j,k) when N(i,j,k) is not specified. However, for subsets with fixed classes of monomials, the computer program forces complete classes into N(i,j,k), thereby biasing the choice of basis towards one which is closed -- towards a basis which is the union of paired and fixed classes.

We conclude this section by extending the concepts of paired
and fixed. An equation e relating monomials of R(i,j,k) is
termed paired when the set of equations obtained by applying the
various t in T(i,j,k) to e consists, after canonicalization, of p
distinct members where p is the number of elements in T(i,j,k).
Note that if t(e) is −e, it is considered a distinct member but,
of course, represents the same relation among the monomials.
Equation e is termed fixed when the set of t(e) consists of p÷2
distinct members after canonicalization. For subsets R(i,j,k)
such that at least two of i, j, and k are equal and nonzero, one
can show that a monomial (or an instance of the LJI) is fixed if
and only if there is a transposition t in T(i,j,k) under which
the monomial (or an instance of the LJI) is fixed.

3.2. Algorithm 1: Selection of Equations Expressing Monomial Dependency

Algorithm 1 is basically a heuristic procedure which given a
monomial m from an ordered set of monomials N(i,j,k), selects an
instance of the LJI or more usually a difference of two instances
of the LJI to express the dependency of the monomial m of
N(i,j,k). The use of the selected instance(s) is restricted
first, to control the magnitude of the integers in the equations
expressing monomial dependency, and second, to obey the requirement (when appropriate) that the basis be closed. As the algorithm proceeds, the restrictions controlling the magnitudes of
the integers present in acceptable equations is gradually relaxed
until they are limited only by the integer representations for
the digital computer.

For the discussion in this section, we recall in Section
2.9.6 that the monomial dependencies for noncanonical spaces are
derived from the corresponding canonical spaces. Therefore, the
discussion of Algorithm 1 is limited to sets of degree type

(i,j,k) with $i \geq j \geq k$. Next, recall that for such sets, the monomials are ordered under g by the images of the second factors. This fact is used to establish certain properties of this algorithm.

Given a monomial in $N(i,j,k)$, Algorithm 1 selects an instance (or a difference of instances) of the LJI by parsing the monomial m into four factors. The factors are assigned to the arguments w, x, y, and z of the LJI in such a manner that one of the six terms in the resulting equation is m itself. There exist several ways of parsing m and assigning the corresponding factors to w, x, y, and z, but only the simplest and most frequently used will be described here.

The choice between the two preferred methods of parsing the monomial depends on the degree of the second factor m_s of m. The first choice applies to monomials m with $d(m_s) > 2$. Define m_{sf} and m_{ss} to be the first and second factors respectively of m_s, and m_{sff} and m_{sfs} to be the first and second factors respectively of m_{sf}. Then we have for the monomial m,

$$m = m_f((m_{sff}m_{sfs})m_{ss}).$$

Now we form the difference of instances

$$q(m_{ss},m_f,m_{sff},m_{sfs}) - q(m_f,m_{ss},m_{sff},m_{sfs}).$$

The first instance is, of course, chosen so that the monomial m appears as its first term, and the second instance is chosen to force, for a particular subset of monomials, the corresponding submatrix of dependency equations to be diagonal. For this difference, first note that the right sides of the two instances cancel, for the right side of the LJI is symmetric in all of w, x, y, and z. This leaves six summands, one of which yields the term m itself. Next, all monomials m' of $R(i,j,k)$ which are produced by the remaining five summands are such that the degree $d(m'_s)$ is less than $d(m_s)$. Hence, the monomial m must be present with a nonzero coefficient in the equation corresponding to the difference under consideration. Finally, the submatrix,

mentioned above, corresponds to that set of monomials m such that
$d(m_s)$ equals 3. This last statement implies that there is a
preference for bases containing monomials with second factor of
degree 1.

The second choice applies to monomials m with $d(m_s) \leq 2$, but
with $d(m_f) > 2$. These monomials m are parsed in the same way as
those above but with m_s and m_f interchanged, yielding a
difference of the form

$$q(m_{fs}, m_s, m_{fff}, m_{ffs}) - q(m_s, m_{fs}, m_{fff}, m_{ffs}).$$

Here, the first term in the first instance yields m in the fac-
tored form

$$m = ((m_{fff}m_{ffs})m_{fs})m_s.$$

However, other terms in either instance may also yield m causing
it to cancel out of the resulting equation. Indeed, the entire
equation is zero when m_s equals m_{fs}.

The rationale behind the preferred ordering in $N(i,j,k)$ can
now be seen by analyzing the first choice for parsing in the
presence of the preferred ordering for $N(i,j,k)$. Since all the
monomials in the equation selected for m have second factors
whose degree is less than that for m except for m itself, then
the only monomials present in (m)'s equation are further down in
the ordering. This assumes the monomials in $R(i,j,k)-N(i,j,k)$
are ordered after those of $N(i,j,k)$. Hence, if we examine the
rectangular coefficient matrix representing the equations developed by the first choice, it is upper triangular with nonzero
diagonal elements. Hence, the equations selected by the first
choice are independent of each other. In addition, they represent,
in the spaces of degree 8 or more, the majority of the equations
expressing the dependencies among the monomials of $R(i,j,k)$.

The nice upper triangular property possessed by the equational system also holds for sets $N(i,j,k)$ with at least two
of i, j, and k nonzero. But, in solving the basis problem for

such an R(i,j,k), there are three additional requirements which
in part affect the use of equations obtained from Algorithm 1.
First, the status (fixed or paired as defined in Section 3.1) of
the monomial and equation must match, which will be discussed
later in the form of an example. Second, the set of dependent
monomials, prior to the application of parsing, must be closed.
Third, the set of equations (quads) which have been considered
(classed as used quads) must be closed. The closure for quads is
analogous to the concept of closure for monomials. The motivation for these requirements is twofold. There is the desire to
have the upper triangular coefficient matrix exhibit the characteristics possessed by the diagonal coefficient matrix. Simultaneously, these additional requirements are most useful in
promoting better behavior of the intermediate arithmetic.

The second choice of parsing, that for monomials m with
$d(m_s) \leq 2$ and $d(m_f) > 2$, does not yield in general equations
whose coefficient matrix has this nice upper triangular property
possessed by the first choice for parsing. Indeed, it often
yields equations that contain monomials occurring earlier than m
in N(i,j,k) whose second factor has degree exceeding 2 and which
must be replaced by substitution to determine whether the equation is independent of those already used to make monomials
dependent. If, after substitution, the resulting equation is
nonzero, then it is independent of those previously used equations and may be used to make m dependent, providing that m still
remains and certain additional properties, such as monomial
singularity, discussed below hold. On the other hand, if the
resulting equation is nonzero but cannot be used to make m
dependent, then it may be used to establish dependency of some
other monomial therein. However, this need to substitute for
dependent monomials may have two unfavorable side effects; first,
the substitution process may increase the magnitude of the
coefficients in the equation to such an extent that later substitution using this equation yields integers so large that they

cannot be tolerated by the digital computer; and secondly, the monomial in question either may vanish after substitution, or may fail to have its status (paired or fixed) match that of the resulting equation.

Because of the first unfavorable aspect of substitution, we reject the use of the resulting equation, when the maximum coefficient in magnitude in the equation exceeds a threshold. The rejected instances, those corresponding to this offensive equation, will eventually be re-examined either by Algorithm 2 described in the next section, by a parsing of another monomial in $N(i,j,k)$, or by again parsing m with Algorithm 1. The first two of these possibilities obviate the need for considering the rejected equation because of linear dependency among the equations. Should the rejected instance be re-examined, the resulting equation, because further monomials may have been made dependent, may yield a completely different equation with smaller coefficients, including the zero equation, thereby avoiding the integer representation problem.

In addition, to retard the growth of the coefficients, we insist that the coefficient of the dependent monomial be a power of 2. Experience shows that this restriction greatly reduces the magnitude of the coefficients in the intermediate calculations. We can prove that this power of 2 restriction propagates at least with regard to the first choice for parsing. In addition this restriction has theoretical significance, as discussed in Section 6.

Partly related to the second unfavorable aspect of substitution, we now discuss both the problem of monomial singularity and the requirement of status-matching for equations and monomials. Suppose e is an equation in a space where t is the only nontrivial transformation in the set $T(i,j,k)$, and suppose further that the paired classes (m_1,m_1') and (m_2,m_2') and the fixed class m_3 appear in e in the following way;

$$e: \quad m_1 + m_1' + m_2 - 4(m_2') + m_3 = 0.$$

The equation e is paired since

$$t(e): \quad m_1 + m_1' - 4(m_2) + m_2' + m_3 = 0$$

is not identical to e. Now consider, to cope with the problem of singularity, the possibility of taking a linear combination of e and t(e) that yields an equation for m_1 in which m_1' is not present. Since the 2×2 coefficient matrix

$$\begin{bmatrix} 1 & 1 \\ 1 & 1 \end{bmatrix}$$

for the pair (m_1, m_1') in the pair of equations e and t(e) is singular, such a combination cannot exist. On the other hand, the combination e + 4t(e) yields an equation for m_2 in which m_2' is not present, since the coefficient matrix

$$\begin{bmatrix} 1 & -4 \\ -4 & 1 \end{bmatrix}$$

for the pair (m_2, m_2') in the pair of equations e and t(e) is non-singular.

To illustrate the status-matching requirement, one might consider using e to make m_1 dependent, regardless of the presence of m_1'. Then t(e) could be used to make m_2 dependent. But then the set of dependent monomials at this point is not closed which violates one of the three additional requirements given above.

Another possibility is that of using the paired equation e to make dependent the fixed monomial m_3. The earlier objection now applies, namely, closing the quad space fails to utilize the information in t(e). Summarizing, if the coefficient matrix for a class of monomials in a class of equations of the same status is nonsingular, then we can readily make that class of monomials dependent upon the remaining monomials in the equations by taking linear combinations of these equations.

The computer program that implements Algorithm 1 uses choices of parsing other than those above and which are briefly

described here. Monomials of degree 4 when both factors have degree 2 are parsed in such a manner that they occur in the term of
the form (wx)(yz) in the instance q(w,x,y,z). Secondly, certain
monomials are parsed in a manner which yields an instance of the
LJI which does not directly mention them, but instead requires
the utilization of information from earlier spaces. These latter
choices of parsing are designed to reduce the amount of substitution in the space under study, thereby further diminishing the
aforementioned difficulties with large integers.

Empirically, we have discovered that the number of fixed and
paired classes of monomials in a closed basis B(i,j,k) is independent of the particular choice of the closed basis. Addition-
ally, the number of fixed and paired classes in the closed basis
matches those numbers of fixed and paired classes of symmetric
elements in the corresponding special Jordan subspace, whenever
the free Jordan subspace is isomorphic to the special Jordan subspace. Whenever the subspaces are not isomorphic these counts do
not match but are only different by one or two in the counts of
each class. Consequently, we further restrict Algorithm 1 by
limiting the number of paired and fixed classes that can be made
dependent.

We now summarize the four restrictions upon the use of an
equation e to make a monomial m dependent, thereby giving the
properties required after substitution -- those properties pro-
mised in the discussion of the second choice of parsing. First,
the status (fixed or paired) of the monomial and equation must
match. Second, the largest coefficient in magnitude must not
exceed some threshold. Third, the coefficient of the monomial
must be a power of two. Fourth and last, the submatrix of the
class of monomials in the equation must be nonsingular.

These restrictions or the choice of N(i,j,k) may lead to the
existence of linearly independent equations which must be considered before the dimension of the subspace V(i,j,k) can be
correctly determined. There are two quite distinct ways to cope

with these equations. First, with the intention of insisting
that the complement of N(i,j,k) be made a basis, the various
restrictions are progressively relaxed. For example, several
passes through N(i,j,k) are made, each time increasing the
threshold for permissible coefficient size. Second, when all
else fails, the program is permitted to make dependent monomials
which are not members of N(i,j,k). In all cases, however, both
the nonsingularity requirement and the powerful requirement of
closure are never relaxed.

Algorithm 1 has one final task, that of continually invoking
Algorithm 2, which is presented in the next section. Each time
Algorithm 1 either uses a quad to make a monomial dependent or
considers a quad whose corresponding equation becomes identically
zero after substitution, an attempt is made to determine addi-
tionally quads which have therefore become linearly dependent.
The mechanism for determining such linear dependence is accomplished by Algorithm 2 which employs quad arithmetic.

3.3. Algorithm 2: Completing the Quad Space

Given a quad (or a certain kind of quad difference), Algorithm 2 determines the quads (or quad differences) that are newly
made linearly dependent. Although other identities are employed
by Algorithm 2, the primary source of linear dependence is the
identity E_0 of Section 5. Algorithm 2 is called whenever a quad
or quad difference is added to the "used quads list". A quad or
quad difference is added to the used quads list when and only
when either the quad or quad difference yields an equation used
for monomial dependency or the quad or quad difference yields a
zero equation. A quad or quad difference yields a zero equation
either because it becomes identically zero during the attempt to
make a monomial dependent or because it has been shown to be
linearly dependent through the use of Algorithm 2.

The adjunction of a single element to the used quads list
may yield, by Algorithm 2, a number of candidate sets S for quad
arithmetic. As a result, Algorithm 2 processes each element
added to the used quads list to thereby generate additional quads
and quad differences as members of candidate sets S. To illustrate how these other quads are generated, consider the following
example. Suppose the difference $(6,4,1,1)-(4,6,1,1)$ is on the
used quads list, and that we are in the process of adding to that
list the quad difference $(4,6,1,1)-(1,6,4,1)$. By transitivity of
addition, the difference $(6,4,1,1)-(1,6,4,1)$ is also available
for processing by quad arithmetic, since

$$(6,4,1,1)-(1,6,4,1) = ((6,4,1,1)-(4,6,1,1)) + \\ ((4,6,1,1)-(1,6,4,1)).$$

If in addition the quad $(1,6,4,1)$ is already on the used quads
list, then, again by transitivity, the quad $(4,6,1,1)$ is now also
available, since

$$(4,6,1,1) = ((4,6,1,1)-(1,6,4,1)) + (1,6,4,1).$$

In this way, Algorithm 2 applies the property of transitivity of
addition of quads to each element added to the used quads list,
and possibly yields, when used in conjunction with those elements
previously appearing on the used quads list, quads or quad differences for the application of quad arithmetic. Also any quad
or quad difference thus shown to be transitively related to
ones already on the list are adjoined to the used quads list.
Similarly, if q and q' are on the list, and if the elements of q'
are a permutation of those of q, then q-q' is also placed on the
list.

Algorithm 2 processes in turn each quad or quad difference
in the following way. If a quad difference is under consideration, let S denote the set of quads including sign therein.
Otherwise, let S be the set consisting of the single quad under
study. Using quad arithmetic, various sets S* which complete S
are generated. Briefly, as discussed in Section 5, the set S*,

containing S, completes the set S if and only if the set S* is
the set of signed quads appearing in a particular instance of E_0.
For each completion S*, the program considers the corresponding
instance of E_0 and determines the number of quads or quad differences therein that are not yet on the used quads list. When that
number is exactly one, the missing quad or quad difference is
known to be linearly dependent and is therefore added to the used
quads list.

Quad arithmetic is the process whereby, for a given set S,
various sets S* are determined. This arithmetic process is
accomplished by using the data structure as follows.

First, if S consists of the quad (q,q_2,q_3,q_4) where $g(m) = q$
and $d(m) > 1$, and if $q = q_f q_s$, then one set S* consists of the
canonicalizations of the seven quads

$$(q,q_2,q_3,q_4), \ (q_2 q_3,q_4,q_f,q_s), \ -(q_4,q_2 q_3,q_f,q_s),$$
$$(q_2 q_4,q_3,q_f,q_s), \ -(q_3,q_2 q_4,q_f,q_s), \ (q_3 q_4,q_2,q_f,q_s),$$
$$-(q_2,q_3 q_4,q_f,q_s).$$

If all of $q_2 q_3$, $q_2 q_4$, and $q_3 q_4$ correspond to basis monomials, S*
is determined. On the other hand, if any of the products, such
as $q_2 q_3$ represent dependent monomials, the corresponding quad
$(q_2 q_3,q_4,q_f,q_s)$ is replaced by the set of quads obtained by replacing $q_2 q_3$ with $g(b_p)$ for each basis monomial b_p in the replacement expression for $q_2 q_3$. For example, if the product

$$u(q_2 q_3) = \sum_{p=1}^{n} u_p \, g(b_p),$$

then the quad $(q_2 q_3,q_4,q_f,q_s)$ is replaced by the set of n quads
$(g(b_p),q_4,q_f,q_s)$ for $p = 1,2,\ldots,n$. The coefficients u and u_p
are ignored since the presence or absence of elements in the used
quads list is of concern. The quad $-(q_4,q_2 q_3,q_f,q_s)$ must also be
replaced by its corresponding set of quads, but the sign must be
propagated throughout the replacement set. Further, if $d(m) > 3$,
then other sets S* are generated by determining all those dependent monomials m' in the space of degree type dt(m) which have the

basis monomial m in the equation expressing m' in terms of basis
monomials. For example, if the non-basis monomial m' is expressed in terms of the basis monomial m (and possibly other
monomials), then S* consists of the canonicalizations of the
seven quads

$$(g(m'),q_2,q_3,q_4), \quad (q_2q_3,q_4,g(m_f'),g(m_s')),$$
$$-(q_4,q_2q_3,g(m_f'),g(m_s')), \quad (q_2q_4,q_3,g(m_f'),g(m_s')),$$
$$-(q_3,q_2q_4,g(m_f'),g(m_s')), \quad (q_3q_4,q_2,g(m_f'),g(m_s')),$$
$$-(q_2,q_3q_4,g(m_f'),g(m_s')),$$

where $m' = m_f' m_s'$. Again, if any products represent dependent monomials, the corresponding quad is replaced by an appropriate set
of quads as described above. Finally, if $d(m) = 1$, we choose not
to form a completion set S*.

When S consists of a difference, say (q,q_2,q_3,q_4) and
$-(q_2,q,q_3,q_4)$, it may be treated in one of several ways depending
upon the degrees of q and q_2. If q and q_2 have degree 1, no set
S* is sought. Since the approach to treating the second quad is
the same as that for the first, we describe the completion of S
to S*, focusing upon the first quad of the difference.

If $d(m) > 1$ where $g(m) = q$, and if $q = q_f q_s$, then one set S*
consists of the canonicalizations of the seven quads

$$(q_3q_4,q_2,q_f,q_s), \quad (q_2q_f,q_s,q_3,q_4), \quad -(q_s,q_2q_f,q_3,q_4),$$
$$(q_2q_s,q_f,q_3,q_4), \quad -(q_f,q_2q_s,q_3,q_4), \quad (q_fq_s,q_2,q_3,q_4),$$
$$-(q_2,q_fq_s,q_3,q_4).$$

Again, if any products represent dependent monomials, the corresponding quad is replaced by an appropriate set of quads as illustrated above. Further, if $d(m) > 3$, other sets S* are generated
by determining all those dependent monomials m' in the space of
degree type $dt(m)$ which have the basis monomial m in their equation. Finally, if $d(m) = 1$, we then focus attention on the
second quad of the difference, and proceed as above. This completes the description of the implemented version of quad
arithmetic.

Whenever quad arithmetic yields an S* with just one quad or quad difference that is not on the used quads list, it is added to the list. It and its derivatives, arising from transitivity, are made available to be analyzed by the quad arithmetic algorithm. This repetitive process can often cascade to yield from one addition to the used quads list hundreds of other quads without resorting to the expansion of a single instance of the LJI. Consequently, hundreds of equational redundancies can be eliminated in this most efficient manner.

However, the selection processes of Algorithm 1 together with the continuous application of Algorithm 2 may not deal with all the relevant quads in a given space R(i,j,k). When not all the quads have been considered in this fashion, those sets S* with just two quads or quad differences absent from the used quads list are visited one at a time. For such sets S*, the equation corresponding to one of the two unused elements is developed. As before, if the resulting equation is independent of those already expressing monomial dependency, it can be used to express another monomial dependency not yet developed and is then added to the used quads list. On the other hand, if it is zero thus showing that it is dependent upon the previous equations, it still is added to the used quads list, and Algorithm 2 proceeds to determine those sets S* with just one element absent from the used quads list. Clearly, there always exists one such S* since the S* just visited which caused the development of an equation will be selected by Algorithm 2. The process continues with all the sets S* with just two elements absent from the used quads list.

We have never applied the technique just given to an S* with three or more absent from the used quads list. Such an S* would require the development of two equations resulting from substitution into the LJI.

4. Closing Functions: The Formal Presentation

In this section we give a formal treatment of closing functions and their properties. We begin with the basic axioms, and then prove a number of lemmas and corollaries which establish the relation between closing functions and "closed" bases, which will be defined in this section. The choice of a closed basis at each step of an iterative approach to the structure problem for the free Jordan ring on 3 letters has a profound effect on the ease with which information is obtained. Each closing function can be said to both well-order the monomials in its domain of definition and canonicalize each of these monomials.

In this treatment we shall use notation and terminology consistent with the earlier material, but we shall attempt to make this section self-contained. In this regard we begin the formalization of the subject matter of Sections 2.5 to 2.8 with the following notation and basic concepts.

Let R denote the free Jordan ring on 3 letters. The elements of R are finite sums of monomials with rational coefficients. Monomials are defined as finite sequences of letters with appropriate grouping symbols, and the choice here, of course, is from among a, b, and c. The ring R is a commutative and nonassociative ring whose elements satisfy the Jordan identity. Of the various forms of this identity, we prefer the linear form given by

$$((yz)w)x + ((xz)w)y + ((xy)w)z = (wx)(yz) + (wy)(xz) + (wz)(xy),$$

and we denote it by LJI.

It is not all of R, however, which directly concerns us in this section, for the focus of attention here is a particular class of functions defined only for certain individual monomials. We therefore need certain concepts defined for monomials alone.

Definition. The *degree type* dt(m) of the monomial m is a triple of integers (i,j,k) which respectively give the number of occurrences in m of a, b, and c.

Definition. The *degree* d(m) of a monomial m of degree type
(i,j,k) is defined as i+j+k.

The monomials in R can then be neatly partitioned into sets
consisting of all monomials of the same degree. Each of these
sets can in turn be partitioned into subsets S(i,j,k), where
S(i,j,k) consists of all monomials m with dt(m) = (i,j,k). For
each S(i,j,k), there exists a vector space V(i,j,k) consisting of
all finite sums of elements from S(i,j,k) with rational coefficients. Then R equals the span of V(i,j,k) for all (i,j,k) with
i+j+k > 0.

Even without the LJI, each of the V(i,j,k) is, of course, finite-dimensional, and the dimension is obviously in general less
than that of its generating set S(i,j,k) because of commutativity
of multiplication. This brings us to the next important concept
for the class of functions to be studied, namely, that of
commutative variant.

Definition. The monomials m and m' are *commutative variants* of
each other if and only if m' can be obtained from m by some number of applications of commutativity of multiplication.

We therefore consider m to be, for example, a commutative
variant of itself. This concept is used in various ways to restrict the domain of each function of the class in question. We
employ therein the obvious concepts of first and second factors
of a monomial m.

Definition. The *first* and *second factors* of a monomial m,
respectively denoted by m_f and m_s, are those monomials such that
$m = m_f m_s$. The monomials a, b, and c, of course, do not possess
first and second factors.

This definition is well-posed because multiplication is nonassociative.

Finally, we need certain definitions which center on the (i,j,k) triples.

Definition. A triple (i,j,k) is called *canonical* if and only if i ≥ j ≥ k, and *noncanonical* if and only if it fails to satisfy this relation.

We also say that a monomial m with dt(m) = (i,j,k) has canonical type if and only if i ≥ j ≥ k. A set S(i,j,k) can also be said to have canonical type. Canonical sets occur in the axiom p_6, which is a symmetry axiom for the class of functions herein.

The (i,j,k) triples also play an important role for certain distinguished subgroups of S_3, the full set of permutations on 3 letters. In this connection, we identify the permutation ab with the permutation 12, abc with 123, and the like.

Definition. We define T(i,j,k) to be the set of t in S_3 which preserve degree type. Therefore, T(i,j,k) is the set of t in S_3 such that t((i,j,k)) = (i,j,k).

Thus, for example, T(2,1,1) = T(5,2,2) is the set consisting of the transposition bc and the identity; T(3,2,1) is just the identity; and, T(2,2,2) is S_3 itself. The subgroups T(i,j,k) do not come into use directly until we give the set of lemmas and corollaries. We can at last turn to the definition of the functions of primary interest.

The class of functions upon which we now concentrate consists of all closing functions, where a closing function g is defined as any function which satisfies the following eight axioms. These axioms are not assumed to be an independent set of axioms. The first of the eight, p_0, can be viewed essentially as initialization; the next three as ordering; the next two restrict the domain of definition and in effect canonicalize; p_6 is an important type of symmetry; and p_7 states that g^{-1} is defined on the positive integers. Obviously, in view of p_0, p_7 is merely present for parallelism with respect to the discussion in Sections 2.5 to 2.8.

$P_0(g)$. The function g is a one-to-one mapping from a subset of the individual monomials in R onto the positive integers such that $g(a) = 1$, $g(b) = 2$, and $g(c) = 3$.

$P_1(g)$. If g is defined on the monomials m and m' with $d(m) < d(m')$, then $g(m) < g(m')$.

$P_2(g)$. If g is defined on m and m' with $d(m) = d(m')$, and if $dt(m) = (i,j,k)$ and $dt(m') = (i',j',k')$, then $g(m) < g(m')$ when either $k < k'$, or $k = k'$ with $j < j'$.

$P_3(g)$. If m and m' are monomials of canonical type on which g is defined, if $dt(m) = dt(m')$, and if g is defined on both the first and second factors of both m and m', then $g(m) < g(m')$ when either $g(m_s) < g(m'_s)$, or $g(m_s) = g(m'_s)$ with $g(m_f) < g(m'_f)$.

$P_4(g)$. If g is defined on m, then $g(m_f) \geq g(m_s)$.

The next property $P_5(g)$ is a canonicalization for distributivity. It is a formalization of the requirement that the monomials in the domain of g are recursively products of basis monomials relative to g.

$P_5(g)$. Let $B(i,j,k)(g)$ be that set of monomials m on which g is defined such that $dt(m) = (i,j,k)$, and such that there exists either an m_1 on which g is defined with m_{1f} identical to m or an m_2 on which g is defined with m_{2s} identical to m. We first require that, for any (i,j,k) such that $B(i,j,k)(g)$ is not empty, $B(i,j,k)(g)$ be precisely a basis for $V(i,j,k)$ (with respect to the LJI). We then require that, if m and m' are respectively members of $B(i,j,k)(g)$ and $B(i',j',k')(g)$ with $g(m) \geq g(m')$, g be defined on mm'. We further require that g defined on m with $d(m) > 1$ implies g is defined on m_f and on m_s.

$P_6(g)$. Let m_1 and m_2 be monomials of canonical type on which g is defined such that $dt(m_1) = dt(m_2)$ and such that

$g(m_1) < g(m_2)$. With transpositions considered simpler than 3-cycles, let t be the simplest permutation on 3 letters such that, for a given noncanonical triple (i',j',k'), $t(m_1)$ and $t(m_2)$ have degree type (i',j',k'). We require of g that there exist monomials m_1' and m_2' such that g is defined on both m_1' and m_2', such that m_1' is a commutative variant of $t(m_1)$ and m_2' of $t(m_2)$, and such that $g(m_1') < g(m_2')$. We also require that, when g is defined on a noncanonical m', there exists a canonical m on which g is defined such that t(m) is a commutative variant of m', where t is the simplest possible permutation.

$p_7(g)$. For every positive integer p, there exists a monomial m on which g is defined such that g(m) = p.

Certain elementary facts can be easily deduced concerning closing functions. Axiom p_0, for example, can be extended to cover the first 102 monomials in the well-ordering. The axiom states that all closing functions agree on the monomials a, b, and c. This agreement actually extends from the monomial a which is mapped to 1 to (cc)(cc) which is mapped to 102 under any function g satisfying p_0 to p_7. All eight axioms contribute to the fact that the domains, ranges, and orderings of such g are identical. Section 2.7 contains some of the details. But at and beyond degree 5, separation can and does occur simply because the LJI applies to the monomials of degree 4.

On the other hand, p_4 forces certain monomials to be in the domain of no function g. For example, b(ba) and (bb)(cc) are mapped by no closing function. There are still other monomials, (((aa)b)a)b for example, for which there exist functions g_1 and g_2 satisfying p_0 to p_7 such that g_1 and g_2 are defined on this m but with $g_1(m) \neq g_2(m)$.

Turning to more significant properties of closing functions, we make the following definition.

<u>Definition</u>. For a given (i,j,k) and a given function g satisfying p_0 to p_7, R(i,j,k)(g) is the set of monomials m on which g is

defined and for which $dt(m) = (i,j,k)$.

One can first show easily that, for all (i,j,k) and all closing functions g, $R(i,j,k)(g)$ is fully canonicalized with respect to commutativity of multiplication. That is, there do not exist monomials m_1 and m_2 with $m_1 \neq m_2$ but with m_2 a commutative variant of m_1 such that m_1 and m_2 are simultaneously members of some $R(i,j,k)(g)$ for some g. In fact one can prove the stronger claim: If g_1 and g_2 satisfy p_0 to p_7, and if m_1 and m_2 are such that g_1 is defined on m_1 and g_2 on m_2, and if m_2 is a commutative variant of m_1, then $m_1 = m_2$. In addition to the canonicalization for commutativity, the $R(i,j,k)(g)$ realize canonicalization for distributivity. For that discussion we need the following definition.

Definition. Let g be a given function for which p_0 to p_7 hold. Then $R(i',j',k')(g)$ is said to be g-*contained* in $R(i,j,k)(g)$ if and only if $i' \leq i$ and $j' \leq j$ and $k' \leq k$ and at least one of the three inequalities is strict.

Now within any given $R(i,j,k)(g)$ for some given closing function g, there are three possible sources of linear dependence among the monomials therein. Commutativity of multiplication is eliminated as a source by the previous remarks. Linear dependence coming from relations among monomials in some $R(i',j',k')(g)$ which is g-contained in $R(i,j,k)(g)$ is eliminated by $p_5(g)$. Thus the only source for relations among the members of $R(i,j,k)(g)$ is the linearized Jordan identity applied directly to $R(i,j,k)(g)$. In other words, one need examine only instances of the LJI arising from substitution of basis elements for w, x, y, z -- elements from $B(i',j',k')(g)$ for relevant (i',j',k'). To be relevant, the $R(i',j',k')(g)$ must each be g-contained in $R(i,j,k)(g)$. The four monomials being substituted for w, x, y, z must be such that the coordinatewise sum of their degree types equals (i,j,k). Finally, to maintain all within the domain of the given g, relations among monomials in the g-contained sets must often be

applied during the evaluation of an instance of the LJI. For
example, the first summand contains the partial product yz which
may evaluate to a monomial which is not a member of the corresponding $B(i',j',k')(g)$. This necessitates a replacement of that
monomial by its equivalent expression involving only members from
the corresponding $B(i',j',k')(g)$. The discussion of this paragraph taken as a whole shows that canonicalization with respect
to distributivity is present in $R(i,j,k)(g)$.

The focus of attention now shifts from the $R(i,j,k)$ to the
$B(i,j,k)$ and the powerful property of closure.

Definition. A subset B_1 of $R(i,j,k)(g)$ is *closed* when, for all t
in $T(i,j,k)$, the monomial m is a member of B_1 if and only if a
commutative variant of $t(m)$ is a member of B_1.

Some familiarity with this concept can be gained by examination of Section 2.8. The concept of closure can be localized in
the obvious way to that of closure with respect to some particular element of $T(i,j,k)$. In this connection we prove the following useful remark.

Remark 1. If a set B_1 is closed under at least two different
transpositions, then B_1 is closed under S_3, the group of all permutations on 3 letters.

Proof. Merely note that any two distinct transpositions generate
S_3, and apply Remark 2.

Remark 2. Let g be any given closing function, and m be any
given monomial on which g is defined. Let t_1, t_2, and t_3 be members of S_3 with $t_1 t_2 = t_3$. It is not assumed that $dt(m) =$
$dt(t_p(m))$ for any p. If m' in $R(i',j',k')(g)$ is a commutative
variant of $t_3(m)$, and if m" in $R(i",j",k")(g)$ is a commutative
variant of $t_2(m)$, then m' is a commutative variant of $t_1(m")$.

Proof. Note that it is canonicalization with respect to $p_4(g)$
which is under consideration. Also note that $t(m_1) = t(m_{1f})t(m_{1s})$
for any m_1 and any permutation t. One is required to apply

commutativity of multiplication some number of times to recursively satisfy $p_4(g)$. The proof is by induction on the number of letters in m. The axioms p_0, p_5, and p_6 are the key.

We now state and prove the most important result for closing functions, the lemma of closed bases for canonical sets. Recall that $p_5(g)$ requires that the $B(i,j,k)(g)$ be bases respectively for $V(i,j,k)$. Therefore, each $B(i,j,k)(g)$ is a basis for $R(i,j,k)(g)$, which permits us to call the $B(i,j,k)(g)$ bases relative to g.

Lemma 1. Let g be any given closing function. Then $R(i,j,k)(g)$ is not empty for all (i,j,k) with $i+j+k > 0$. More importantly, if $R(i,j,k)(g)$ is such that $i \geq j \geq k$, and if $B(i,j,k)(g)$ is that set defined in $p_5(g)$, then $B(i,j,k)(g)$ is closed and a basis for $R(i,j,k)(g)$.

Proof. For the first half of the lemma, assume by way of contradiction that $R(i,j,k)$ is empty with $i+j+k > 0$ but that no earlier, within the ordering dictated by p_1 to p_3, $R(i',j',k')$ is empty. By p_0, $R(i,j,k)$ is of degree at least 2. Assume without loss of generality that $i > 0$. Then $R(i-1,j,k)$ is defined and not empty since $R(i,j,k)$ is assumed to be the earliest empty set contradicting the lemma and, by p_1, $R(i-1,j,k)$ is earlier in the ordering. Let m be a basis monomial in the basis underlying g in $R(i-1,j,k)$. The monomial a is a basis monomial in $(1,0,0)$. By one of p_1 to p_3, $g(m) \geq g(a)$. Thus the monomial ma satisfies the restriction of p_4. Therefore, by p_5, g must be defined on ma, which contradicts the assumption that $R(i,j,k)(g)$ is empty.

For the second half of the lemma, assume again by way of contradiction that there exists an $R(i,j,k)(g)$ falsifying the second part of the lemma. For the type $(i,0,0)$, the argument is trivial in that the only relevant permutation is the identity. Without loss of generality assume that $i = j \geq k$. Then the basis relevant to g for this $R(i,j,k)$ is not closed. By Remark 1, there must exist a transposition, therefore, under which the basis is not closed. Assume without loss of generality that the transposition is ab. There then exists a monomial m such that m is in the non-closed basis of this $R(i,j,k)$ while the canonicalization with respect to p_4 of ab applied to m is not in the basis. Let t denote the transposition ab. By p_5, g must be defined on the monomial mb. By p_6, g is defined on the canonicalization with respect to p_4 of t(mb), which is equal to the canonicalization of t(m) multiplied on the right by the monomial a. By p_5, the canonicalization of t(m) must be in the basis of $R(i,j,k)(g)$ upon which g rests, which is a contradiction.

Lemma 1 can be extended to include the sets of noncanonical type. To do so, we must first prove Lemma 2 which exhibits the control $p_6(g)$ exercises over those noncanonical sets. By definition, p_6 already forces an order on the elements of each non-canonical $R(i',j',k')(g)$ determined by that in the appropriate canonical $R(i,j,k)(g)$, which is why those noncanonical sets are called permutation images of the canonical.

Lemma 2. Let g be any closing function, and let $R(i,j,k)(g)$ be given such that $R(i,j,k)(g)$ is canonical. If t in S_3 is such that $t(R(i,j,k)(g)) = R(i',j',k')(g)$ is noncanonical, then the canonicalization with respect to commutativity of $t(B(i,j,k)(g))$ is a basis for $R(i',j',k')(g)$ and in fact is equal to $B(i',j',k')(g)$.

Proof. If $(i,j,k) = (1,0,0)$, p_0 is sufficient to establish the result. So assume the degree of m is greater than 1 for m in $B(i,j,k)$. By p_5, the monomials ma are in the domain of definition of g in $R(i+1,j,k)$. These monomials are obviously canonicalized with respect to p_4 since all m in $B(i,j,k)$ are. Let t be a permutation on 3 letters with $t(R(i,j,k)) = R(i',j',k')$ and with $i' \geq j' \geq k'$ false. Then by p_6 the canonicalization of t(ma) for m in $B(i,j,k)$ must be in the domain of definition of g in $R(i',j',k')$. For any such ma, the canonicalization of t(ma) is just the product of the canonicalization of t(m) with t(a).

Now p_5 then implies that the canonicalization of $t(m)$ is in the basis relevant to g of $R(i',j',k')$. Thus the images under t of the basis monomials of $R(i,j,k)$ are basis monomials of $R(i',j',k')$. Since $R(i,j,k)$ and $R(i',j',k')$ are isomorphic and thus have the same dimension, the proof is complete.

Before presenting the extension of Lemma 1, we give Lemma 3 which establishes a type of closure for all $R(i,j,k)(g)$.

Lemma 3. Let g be any closing function, and let $R(i,j,k)(g)$ be given for some (i,j,k). If m is in $R(i,j,k)(g)$ and t is in S_3 with $dt(m) = dt(t(m))$, then the canonicalization with respect to commutativity of $t(m)$ is in $R(i,j,k)(g)$.

Proof. If m has degree 1, there is nothing to prove. Assume by way of contradiction that the lemma is false, and let, therefore, $R(i,j,k)$ be the earliest set with respect to the ordering corresponding to g which contradicts the lemma. Assume that m is in $R(i,j,k)(g)$, that the canonicalization with respect to p_4 of $t(m)$ is not in $R(i,j,k)$ but has degree type (i,j,k), and that the degree of m is greater than 1. Let $m = m_f m_s$, the product of its first and second components. The canonicalization of $t(m)$ is either the product of the canonicalization of $t(m_f)$ multiplied on the right by the canonicalization of $t(m_s)$ or the product taken in the other order. Assume without loss of generality that it is the former. By p_5, m_f and m_s are in their respective bases relevant to g.

Case 1. The monomial m_f has degree type (i',j',k') with at least two of i', j', and k' equal, and $t(m_f)$ also has degree type (i',j',k'). By p_1, $R(i',j',k')$ is earlier in the ordering dictated by g than is $R(i,j,k)$. Therefore, by the assumption that $R(i,j,k)$ is the earliest set contradicting the lemma, the canonicalization with respect to p_4 of $t(m_f)$ must be in $R(i',j',k')$. Of course, $i' \geq j' \geq k'$ may be true or false. In either case, since the product of two permutations on 3 letters is also one, we can apply Lemma 1 or Lemma 2 to conclude that the canonicalization of $t(m_f)$ with respect to p_4 is a basis monomial.

Case 2. The monomial m_f has degree type (i',j',k'), but $t(m_f)$ does not. No assumption is made for this case about inequalities among i', j', and k'. If $i' \geq j' \geq k'$, then Lemma 2 requires the canonicalization of $t(m_f)$ to be a basis element. If it is not the case that $i' \geq j' \geq k'$, then let t_1 be a permutation on 3 letters such that the degree type of $t_1(m_f) = (i'',j'',k'')$ with $i'' \geq j'' \geq k''$. The canonicalization of $t_1(m_f)$ can be seen to be a basis element by noting two facts. First, all permutations on 3 letters are one-to-one on the set of all monomials. Second, no $R(i,j,k)(g)$ of any closing function g can contain more than one representative of an equivalence class of commutative variants of a given monomial. Now again we have the proof that the canonicalization of $t(m_f)$ is a basis monomial from Lemma 2 and the fact that the product of two permutations on 3 letters is also one, which completes Case 2.

In a similar manner, we conclude that the canonicalization with respect to p_4 of $t(m_s)$ is a basis monomial. Therefore, by p_5, the product of the canonicalizations of $t(m_f)$ and of $t(m_s)$ is in $R(i,j,k)$. But this contradicts the assumption on m, and the proof is complete.

We now extend Lemma 1 to establish the uniform existence of closed bases.

Corollary 1. The bases $B(i,j,k)(g)$ are closed for all closing functions g and all (i,j,k).

Proof. Let g be some given closing function. If (i,j,k) is such that no two of i, j, and k are equal, then the identity is the only relevant permutation to consider for closure, and there is nothing to prove. If $i \geq j \geq k$ with at least two of i, j, and k equal, Lemma 1 completes the proof. The last case is that for which exactly two of i, j, and k are equal, but $i \geq j \geq k$ is

false. Assume, therefore, without loss of generality that
$i < j = k$. There then exists an $R(i',j',k')(g)$ with $i' = j' > k'$
such that $t'(R(i',j',k')) = R(i,j,k)$, where t' is the transposition ac. By Lemma 2, the basis $B(i,j,k)(g)$ consists of the
canonicalizations of $t'(m')$, where m' is in $B(i',j',k')(g)$. Now
let m be a basis monomial in $B(i,j,k)$. We must show for closure
that the canonicalization with respect to p_4 of $t(m)$ is in
$B(i,j,k)$, where t is the transposition bc. All other permutations on 3 letters other than the identity are excluded from
$T(i,j,k)$ since they fail to map $R(i,j,k)$ to a set with the degree
type (i,j,k). By Lemma 2, there is a basis monomial m' in
$R(i',j',k')$ such that, after canonicalization with respect to p_4,
$t'(m')$ is m. After noting that tt' is the 3-cycle abc, we apply
Remark 2 and find that the canonicalization of $t(m)$ is that monomial obtained by the canonicalization of $t_3(m')$, where t_3 is abc.
Now let t'' be the transposition ab, and note that $t_3 = t't''$ in
preparation for a second application of Remark 2. We thus obtain
the fact that the canonicalization of $t(m)$ is the monomial obtained from the canonicalization of t' applied to the canonicalization of $t''(m')$. But, by Lemma 1, the canonicalization of
$t''(m')$ is in $B(i',j',k')(g)$. Therefore, by Lemma 2, the
canonicalization of t' applied to the canonicalization of $t''(m')$
is in $B(i,j,k)$. Since m was arbitrarily chosen, the proof is
complete.

Finally, we have the result which is explicitly used in the
inverted approach of Section 2.4 -- our approach to obtaining basis information for subspaces of the free Jordan ring on 3
letters.

Corollary 2. For any given closing function g and any given
(i,j,k), the set of monomials in $R(i,j,k)(g)-B(i,j,k)(g)$ is
closed under $T(i,j,k)$.

Proof. Note that t^{-1} is t itself when t is a transposition, and
t^{-1} is tt when t is a 3-cycle. The proof is one by contradiction.

Assume m is in $R(i,j,k)(g)$ and is non-basis, while the canonicalization of $t(m)$ is a basis monomial, where t is some permutation
such that the degree type of $t(R(i,j,k))$ is (i,j,k). Now apply
Corollary 1, Lemma 3, and Remark 2 to obtain the fact that m itself is in $B(i,j,k)$, which is the contradiction.

5. Quad Arithmetic and the Identity E_0

In this section we give an introduction to quad arithmetic
-- an arithmetic which plays an essential role in our attack on
the structure problem for the free Jordan ring on 3 letters. The
arithmetic is based on a surprisingly powerful identity, whose
validity is proven in this section, which establishes relations
among certain instances of the linearized Jordan identity. Recall that the approach herein to the structure problem takes the
form of a set of subproblems, each of which requires the finding
of a basis for a given $V(i,j,k)$, where $V(i,j,k)$ is the space
spanned by the set of all monomials in R of type (i,j,k). In
this inverted or unnatural approach, for which an overview is
found in Section 2.4 and a detailed account in Section 3, each
subproblem of $V(i,j,k)$ is replaced by one for $R(i,j,k)(g)$ for
some closing function g. Although this replacement avoids,
through reliance on closing functions and their properties, a
number of the difficulties which may be encountered in basis
finding, there still remains the serious problem of equational
redundancy. Quad arithmetic is the mechanism which virtually
eliminates the seriousness of this problem.

Before turning to the treatment of quad arithmetic, the problem of redundancy as it affects basis finding must be clearly
understood. In the previous section, following the definition of
"g-contained", one sees that all of the linear dependency among
the monomials of $R(i,j,k)(g)$, for some given (i,j,k) and some
given g, can be found by examining a restricted set of instances

of the LJI. For the given $R(i,j,k)(g)$, one need only consider
instances arising from the substitution of basis elements m_1, m_2,
m_3, m_4 respectively for w, x, y, z, where the coordinatewise sum
of $dt(m_p)$ with $p = 1,2,3,4$ equals (i,j,k), and where each of m_1
to m_4 is a member of some $B(i',j',k')(g)$. The implicit assumption here is, of course, that g is not actually a closing function but is in fact an initial segment of one. Thus, bases for
$R(i',j',k')(g)$ have been found for the sets g-contained in
$R(i,j,k)(g)$ having properties consistent with those required by
closing functions. The object of the game, therefore, is to extend g, by finding an appropriate $B(i,j,k)(g)$, in a fashion
consistent with P_0 to P_7.

The finding of a suitable $B(i,j,k)(g)$, one which is a closed
subset of $R(i,j,k)(g)$, is made substantially easier by the elimination through P_4 and P_5 both of commutative variants and of two
important classes of redundant equations. A redundant equation
is one which is linearly dependent on others present in the system -- one whose contained information thus duplicates information available elsewhere. The first class to be eliminated is
the set corresponding to instances of the LJI coming from a
substitution for w, x, y, z in which not all of the m_p are basis
monomials relative to g. The second class consists of the relations coming from the g-contained sets which could then be
appropriately multiplied by some monomial to make them possibly
relevant to $R(i,j,k)(g)$.

In addition to this canonicalization for commutativity and
distributivity, one further canonicalization, that for symmetry
in the LJI, can be made to thus eliminate another class of redundant equations. This redundancy is illustrated by taking m_1 to
m_4 as one set of four monomials to be respectively substituted
for w, x, y, z in the LJI, and taking as a different set m_1 with
m_2' to m_4', where m_2' to m_4' are simply a permutation of m_2 to m_4.
The corresponding two instances of the LJI yield identical linear
equations after application of commutativity of addition and

multiplication. This redundancy, coming from the symmetry in x, y, and z in the LJI, can be immediately eliminated by an ordering of monomials such as that which occurs with the use of closing functions or initial segments of closing functions.

Unfortunately, even after elimination of the various types of redundancy described above, there is still present much linear dependence among the remaining linear equations corresponding to the instances of the LJI, at least for the interesting $R(i,j,k)(g)$. In other words, for those interesting $R(i,j,k)$, the matrix of remaining instances of the LJI is far from full rank. It is this phenomenon which motivates the interest in quad arithmetic. In this connection, we shall continue to refer to the unneeded equations as redundant equations to avoid conflict with the phrase "linearly dependent" occurring in the discussion of dependency among members of $R(i,j,k)(g)$. It is through quad arithmetic that the vast majority of these remaining redundant equations are identified and thus eliminated from consideration and without recourse either to substitution into the LJI or to any of the techniques for matrix manipulation.

We now turn to the main discussion of quad arithmetic. For the remainder of this section let g denote either a closing function or an initial segment thereof.

Definition. For a given closing function g or an initial segment of a closing function g, a *quad* $q(g)$ is an ordered set (q_1, q_2, q_3, q_4) of four positive integers such that there exists four monomials m_1 to m_4 in the domain of g with respectively $q_p = g(m_p)$, for $p = 1,2,3,4$. Put differently, a quad $q(g)$ is an ordered set of four ordered pairs $m_1, g(m_1), \ldots, m_4, g(m_4)$.

Definition. The *degree type* $dt(q(g))$ is the coordinatewise sum of $dt(m_p)$ for $p = 1,2,3,4$, where $q(g) = (q_1, q_2, q_3, q_4)$ and $g(m_p) = q_p$ with $p = 1,2,3,4$.

Definition. A quad $q(g) = (q_1, q_2, q_3, q_4)$ is *canonical* if and only if both $q_2 \geq q_3 \geq q_4$ and m_p for $p = 1,2,3,4$ is a member of some

$B(i',j',k')(g)$.

Definition. The set $QR(i,j,k)(g)$ for some (i,j,k) and some g consists of all $q(g)$ such that $q(g)$ is both canonical and of degree type (i,j,k).

For each function g, there exists a companion function g* which maps the set of quads $q(g)$ into the set of instances of the LJI. The function g* maps $q(g)$ to the substitution instance of the LJI corresponding to the substitution of m_1 for w, m_2 for x, m_3 for y, and m_4 for z, where m_p with $p = 1,2,3,4$ is such that $g(m_p) = q_p$ with $q(g) = (q_1,q_2,q_3,q_4)$. The mapping g* is clearly one-to-one. (Note that an instance of the LJI is not simply the equation obtained therein but is instead the equation coupled with its history. Thus, for example, an interchange of the values to be substituted for y and z yields two distinct instances of the LJI, when the values are unequal, even though the same linear equation is produced.)

The most significant property immediately obtainable for g* is that g* maps $QR(i,j,k)(g)$, for any given (i,j,k) and any given g, to that set of instances of the LJI which determine the dependency relations among the monomials of $R(i,j,k)(g)$ -- those instances which remain after the elimination of the various classes of trivially redundant equations. As we remarked earlier, there exist among the equations corresponding to these remaining instances a large number of relations. Many of these relations are instances of a single identity E_0. The identity E_0 can be said to act on instances of the LJI in a manner similar to that of the LJI itself on monomials.

To understand the use of the identity E_0, first note that the simplest application of the LJI can be generally described as follows. The LJI yields relations among the individual monomials where the monomials in any such relation possess certain characteristics in common. Before any additional substitution, the monomials in such a relation number six. They all have the same

degree type. Their first and second factors are each determined
by the assignment respectively to w, to x, to y, and to z. For
this simple application, one can summarize by stating that the
linearized Jordan identity is an identity in four variables
which relates six individual monomials. In the same spirit, E_0
can be described as an identity in five variables which relates
seven instances of the LJI. All seven instances have the same
degree type, which makes E_0 potentially useful in identifying
occurrences of redundancy among sets of equations yielded by the
LJI and being considered for a given $R(i,j,k)(g)$.

In general let $q(m_1, m_2, m_3, m_4)$ be the expression,

$$((m_3 m_4) m_1) m_2 + ((m_2 m_4) m_1) m_3 + ((m_2 m_3) m_1) m_4 - (m_1 m_2)(m_3 m_4)$$
$$- (m_1 m_3)(m_2 m_4) - (m_1 m_4)(m_2 m_3),$$

where m_1, m_2, m_3, and m_4 are four given monomials. Clearly,
$q(m_1, m_2, m_3, m_4)$ is closely related to an instance of the LJI, and
in fact, by setting $q(m_1, m_2, m_3, m_4)$ equal to zero, one obtains
that instance which is yielded by substituting m_1 for w, m_2 for
x, m_3 for y, and m_4 for z. We therefore, as in Section 3, often
identify the expression $q(m_1, m_2, m_3, m_4)$ with the corresponding instance of the LJI.

The identity E_0 is:

$$q(m_u m_v, m_x, m_y, m_z) =$$
$$q(m_y m_z, m_x, m_u, m_v) - q(m_x, m_y m_z, m_u, m_v) +$$
$$q(m_x m_z, m_y, m_u, m_v) - q(m_y, m_x m_z, m_u, m_v) +$$
$$q(m_x m_y, m_z, m_u, m_v) - q(m_z, m_x m_y, m_u, m_v).$$

First note that in E_0 no assumption is made concerning the
basis status of any of the arguments of any of the q's therein.
Next note that the left side of the identity corresponds to an
instance of the LJI in which the monomial to be substituted for w
has degree greater than or equal to 2. Then note that the six
expressions on the right of E_0 naturally group into three differences. Each difference has the property that the two expressions

therein have three terms in common but opposite in sign, which
therefore cancel. These common terms occur because of the
symmetry in w, x, y, and z on the right side of the LJI. These
remarks now bring us to the proof of E_0.

<u>Proof</u>. The left side of E_0 evaluates to

$$((m_y m_z)(m_u m_v))m_x + ((m_x m_z)(m_u m_v))m_y + ((m_x m_y)(m_u m_v))m_z -$$
$$((m_u m_v)m_x)(m_y m_z) - ((m_u m_v)m_y)(m_x m_z) - ((m_u m_v)m_z)(m_x m_y).$$

Denote this expression by $E(q_0)$. In a similar manner, each of
the six expressions on the right side can be evaluated. After
applying commutativity of both addition and multiplication, can-
celling like terms, and collecting, the right side evaluates to

$$E(q_0) + m_u E(q_1) + m_v E(q_2),$$

where $E(q_1)$ and $E(q_2)$ denote expressions which we will show evaluate to 0. The first expression on the right side of E_0 contributes to $E(q_1)$ the term $(m_v m_x)(m_y m_z)$ and to $E(q_2)$ the term
$(m_u m_x)(m_y m_z)$. The second expression on the right contributes
$-(((m_y m_z)m_v)m_x)$ to $E(q_1)$ and $-(((m_y m_z)m_u)m_x)$ to $E(q_2)$. By com-
pleting the computation for the contributions to $E(q_1)$ from the
third through the sixth expressions on the right side of E_0, we
find that $E(q_1)$ is identical to the expression $-q(m_v, m_x, m_y, m_z)$.
Thus $E(q_1)$ is zero since subtracting the left side from the right
of an instance of the LJI is zero. Similarly, $E(q_2)$ equals zero
since the right side minus the left side of the equation corresponding to $-q(m_u, m_x, m_y, m_z)$ equals zero. The proof is complete.

The value of E_0 depends on both the ease of application and
the frequency with which it establishes redundancy among equations. To apply E_0, one need merely factor certain monomials and
multiply certain monomials. The frequency question, on the other
hand, is more complex to answer in that it depends on the particular choice within $R(i,j,k)(g)$ of the set $N(i,j,k)(g)$ -- that set
whose complement in $R(i,j,k)(g)$ will hopefully turn out to be a
basis -- and on the approach employed to finding the dependency

relations among the members of R(i,j,k)(g). The identity E_0 very
frequently yields information for us because of the use of Algorithm 1, given in Section 3, and because of the characteristics
we often impose on the choice of basis (see Section 6). The
identity E_0 is in fact the basis for the quad arithmetic employed
in Algorithm 2, also found in Section 3.

Quad arithmetic modulo both the identity E_0 and the function
g is that procedure which finds, for any given set S of signed
quads q(g), all sets S* such that S is contained in S* and such
that the elements of S* all occur in a single instance of E_0 and
with the correct sign. If, for example, S_1 consists of the quads
(4,2,2,1) and -(1,4,2,2), then S_2 will complete S_1 to an instance
of E_0 by adjoining the quad (6,1,1,1). Quad arithmetic finds
that S_2 is an S* for S_1 and is the only such set. If the set S
of quads for consideration contains too many members, for
example, quad arithmetic yields no S* of the appropriate type.
Thus, if the quad difference (4,2,2,1) - (1,4,2,2) is used to
yield a dependency relation in R(3,2,0), then the quad (6,1,1,1)
is redundant -- the equation corresponding to the instance of the
LJI will yield no additional information, where the instance is
the correspondent of q(bb,a,a,a). Here the function g* comes
into play. This last example is made more understandable by use
of the table at the beginning of Section 2.5 and by examination
of the algorithms of Section 3.

This last example typifies the ideal outcome for the application of quad arithmetic. When a set S* completes S by the
adjunction of a single quad or a single quad difference, and when
all of S has already been accounted for as used, then the adjunction to S can be marked as redundant. Quads or differences may
be termed used, for example, either by yielding an equation
expressing monomial dependency or by being marked as redundant by
a previous application of quad arithmetic. What can and does
often happen is that a quad difference yields a dependency relation among monomials, which causes some other quad to be now

known to be redundant by quad arithmetic, which in turn marks yet
a new quad as redundant, ..., which cascades for as many as 200
quads before coming to a halt.

Thus g*, quad arithmetic, E_0, Algorithms 1 and 2, and the
preferred class of bases combine to find the nontrivial occurrences of redundant equations encountered in finding bases for
$R(i,j,k)(g)$. The choice of the function g and of the identities
upon which quad arithmetic is based can and do have a sharp
effect on its usefulness.

6. Dimension, the Kernel, Conjectures, and Open Questions

In this section we present the results obtained by use of
the inverted or unnatural approach outlined in Section 2.4. That
approach, based on the closing functions of Section 4 and quad
arithmetic of Section 5, yielded answers to various open questions with surprising ease when compared to a straightforward
approach. Although the results are given in terms of concepts
appearing earlier in the paper, we shall supply at appropriate
points a brief but hopefully adequate hint to convey the corresponding meaning in more familiar terms. One can therefore glean
from this section alone the facts concerning the free Jordan ring
R on 3 letters -- the facts obtained by the study upon which this
paper is based. On the other hand, if the intention is to apply
the techniques herein to problems in other algebras by, for example, designing a computer program similar to that employed here,
one must concentrate on Sections 2 and 3.

6.1. Results: Dimension, Bases, and the Kernel

We begin with Table 2 which, for all subspaces of degree
less than or equal to 9 of R, gives the dimension, the rank, the
number of equations including those which are nontrivially

redundant, the dimension of the kernel, and the time required to obtain the information. An equation is considered nontrivially redundant if it cannot be discarded either by canonicalization for commutativity of multiplication, or by canonicalization for distributivity, or by canonicalization for the symmetry in the LJI with respect to x, y, and z. To fully utilize this table, we recall the following. The subspaces of R which are of natural concern are $V(i,j,k)$, where $V(i,j,k)$ consists of all finite sums with rational coefficients of all individual monomials of degree type (i,j,k). Each $V(i,j,k)$ is studied by instead studying $R(i,j,k)(g)$ for some closing function g. Each $R(i,j,k)(g)$ is a subset of the corresponding $V(i,j,k)$ consisting of individual monomials of degree type (i,j,k) for which canonicalization both with respect to commutativity of multiplication and to distributivity have been applied. The dependency of $R(i,j,k)(g)$ on a given closing function g reflects the fact that bases have already been chosen for so-called earlier subspaces. In particular, the elements of a given $R(i,j,k)(g)$ are recursively products of monomials from already chosen bases. Since a basis for a given $R(i,j,k)(g)$ is clearly one for the corresponding $V(i,j,k)$, the dimension given in the table is that for either.

Next under discussion is the column labeled rank. On the one hand, for a given (i,j,k), the rank plus the dimension gives the number of elements in $R(i,j,k)(g)$. Although the functions g automate the various types of canonicalization, the particular choice therein has no effect on the cardinality of the $R(i,j,k)$. On the other hand, the rank gives the number of linearly independent equations which must be found within the subspace under study -- the equations expressing the dependent monomials solely in terms of the independent basis monomials. These equations are used to solve the basis problem for later spaces. For example, an iterative approach to the problem in which $R(3,3,2)$ is the current target relies on, among others, the dependency relations from R(3,2,2), R(3,2,1), and R(2,1,1). When the rank is considered in conjunction with the next object of discussion, the number of equations (quads), the seriousness of the problem of nontrivial equational redundancy becomes evident.

An equation is nontrivially redundant if two properties hold for it. First, it must remain after the discarding of equations because of canonicalization for commutativity of multiplication, or canonicalization for distributivity, or canonicalization for symmetry in x, y, and z in the linearized Jordan identity. Second, it must be linearly dependent on others which remain after the canonicalizations just given. The number of remaining equations including those which are nontrivially redundant is that which is given in the table. This number, therefore, gives the number of quads, or equivalently the number of instances of the LJI, which must be considered in seeking a basis for a given R(i,j,k)(g). The relevant instances of the LJI are just those which are obtained by substitution for w, x, y, and z of only basis monomials whose degree types are commensurate with (i,j,k) and, of course, exluding those yielding obviously duplicate equations because of the x, y, z symmetry. Thus, from the viewpoint of linear algebra, the matrix from which the dependency relations among monomials are to be obtained is a q×p matrix, where q is the equational number in the table and p the sum there of the dimension and rank.

For the next table entry, kernel dimension, recall the fact that there is a homomorphic mapping from the free Jordan ring R to the special Jordan ring. The kernel is that set of elements which is mapped to 0. It is well-known [2] that, if at least one of (i,j,k) is less than or equal to 1, then the subspace V(i,j,k) is mapped isomorphically to the corresponding subspace in the special Jordan ring, and hence the dimension of the kernel in R(i,j,k) equals 0 when at least one of the triple is less than or equal to 1. The dimension in R(i,j,k)(g) of the kernel is the number of linearly independent elements which span the kernel

Table 2

Degree Type	Dimension	Rank	Number of Equations(Quads)	Kernel Dimension	Time (Sec.)
(1,0,0)	1	0	0	0	0
(2,0,0)	1	0	0	0	0
(1,1,0)	1	0	0	0	0
(3,0,0)	1	0	0	0	0
(2,1,0)	2	0	0	0	0
(1,1,1)	3	0	0	0	0
(4,0,0)	1	1	1	0	0
(3,1,0)	2	2	2	0	0
(2,2,0)	4	2	2	0	0
(2,1,1)	6	3	3	0	0
(5,0,0)	1	1	2	0	0
(4,1,0)	3	3	5	0	0
(3,2,0)	6	5	8	0	0
(3,1,1)	10	8	12	0	0
(2,2,1)	16	11	16	0	0
(6,0,0)	1	2	4	0	0
(5,1,0)	3	6	13	0	0
(4,2,0)	9	12	24	0	0
(3,3,0)	10	15	29	0	0
(4,1,1)	15	19	39	0	0
(3,2,1)	30	34	64	0	1
(2,2,2)	48	48	87	0	1
(7,0,0)	1	2	7	0	0
(6,1,0)	4	8	25	0	0
(5,2,0)	12	20	56	0	1
(4,3,0)	19	30	80	0	1
(5,1,1)	21	34	93	0	1
(4,2,1)	54	76	189	0	3
(3,3,1)	70	96	232	0	3
(3,2,2)	108	140	326	0	6

Table 2 (cont'd)

Degree Type	Dimension	Rank	Number of Equations(Quads)	Kernel Dimension	Time (Sec.)
(8,0,0)	1	3	11	0	0
(7,1,0)	4	12	46	0	0
(6,2,0)	16	34	118	0	1
(5,3,0)	28	60	196	0	2
(4,4,0)	38	73	232	0	4
(6,1,1)	28	58	201	0	2
(5,2,1)	84	154	480	0	8
(4,3,1)	140	240	713	0	18
(4,2,2)	216	352	1011	0	60
(3,3,2)	281	451	1266	1	115
(9,0,0)	1	3	16	0	0
(8,1,0)	5	15	75	0	1
(7,2,0)	20	50	221	0	2
(6,3,0)	44	102	420	0	6
(5,4,0)	66	146	576	0	9
(7,1,1)	36	88	382	0	4
(6,2,1)	128	274	1063	0	21
(5,3,1)	252	508	1850	0	79
(4,4,1)	318	626	2224	0	94
(5,2,2)	384	750	2654	0	139
(4,3,2)	638	1193	4037	2	810
(3,3,3)	846	1545	5118	6	291

therein. The elements are clearly sums of monomials.

The final entry gives the time in seconds on an IBM 370/195 required by the program to find a basis for the subspace under study.

A glance at Table 2 shows that the subspace of type (3,3,2) is the earliest which is not isomorphically mappable to the special Jordan ring. But, in view of Glennie's identity of degree 8 [1], what is new and interesting is the fact that the kernel in type (3,3,2) has dimension 1. Of added interest is the dimension of the kernel at type (4,3,2) and type (3,3,3). In type (4,3,2), two independent kernel elements exist -- the one, of course, is a direct descendant of the kernel element from type (3,3,2) while the other is "new". In type (3,3,3), there are three new kernel elements in addition to the three which are descended from (3,3,2) and its permutation images of type (3,2,3) and of type (2,3,3).

The representation of, say, the (3,3,2) kernel element, of course, depends on the choice of bases in a number of subspaces. Our preference among closing functions g, and hence among bases, has the (3,3,2) kernel element in the form

$$a_1(m_1-m_1') + a_2(m_2-m_2') + \ldots + a_{34}(m_{34}-m_{34}'),$$

where m_p and m_p' with $p = 1,2,\ldots,34$ are individual monomials in $B(3,3,2)(g)$ with m_p' equal to the transposition ab applied to m_p. The coefficients $a_1,a_2,\ldots,a_5$ equal 1; $a_6,a_7,\ldots,a_{18}$ equal 2; $a_{19},a_{20},\ldots,a_{30}$ equal 4; a_{31},a_{32},a_{33} equal 8; and a_{34} equals 12. The monomials are:

$$m_1 = ((((bb)(aa))(cc))b)a \qquad m_2 = (((((bb)a)b)(aa))c)c$$
$$m_3 = (((((aa)a)b)(bb))c)c \qquad m_4 = (((((aa)b)a)c)c)(bb)$$
$$m_5 = (((((bb)b)a)c)c)(aa)$$
$$m_6 = (((((aa)c)c)(bb))b)a \qquad m_7 = ((((cc)(bb))(aa))a)b$$
$$m_8 = (((((aa)c)c)(ba))b)b \qquad m_9 = ((((cc)(bb))(ba))a)a$$
$$m_{10} = ((((bb)(aa))(cb))a)c \qquad m_{11} = (((((cc)b)a)b)b)(aa)$$
$$m_{12} = (((((bb)c)a)c)b)(aa) \qquad m_{13} = ((((cb)(aa))c)a)(bb)$$

$$m_{14} = (((((cc)a)a)b)a)(bb) \qquad m_{15} = ((((bb)c)c)a)b)(aa)$$
$$m_{16} = ((((cc)(aa))b)a)(bb) \qquad m_{17} = (((((bb)c)c)a)a)(ba)$$
$$m_{18} = ((((cc)(aa))b)b)(ba)$$
$$m_{19} = ((((((cb)a)a)b)b)a)c \qquad m_{20} = ((((((ca)a)b)b)b)a)c$$
$$m_{21} = ((((((ba)a)c)b)b)a)c \qquad m_{22} = ((((((ca)(bb))a)a)b)c$$
$$m_{23} = ((((((bb)c)a)a)b)a)c \qquad m_{24} = ((((((bb)a)c)a)b)a)c$$
$$m_{25} = ((((((aa)b)b)c)b)a)c \qquad m_{26} = (((((ba)(ba))c)a)b)c$$
$$m_{27} = (((((ca)(bb))b)a)a)c \qquad m_{28} = (((((bb)a)a)(ca))b)c$$
$$m_{29} = (((((aa)b)b)(ca))b)c \qquad m_{30} = ((((ba)(ba))(cb))a)c$$
$$m_{31} = ((((((cb)a)a)b)a)b)c \qquad m_{32} = ((((((bb)a)a)c)b)a)c$$
$$m_{33} = (((((bb)(aa))c)b)a)c$$
$$m_{34} = (((((cb)(aa))b)a)b)c$$

The choice of bases has an impressive effect on the accompanying arithmetic. A poor choice, for example, in R(3,3,2) is accompanied by such badly behaved arithmetic, such as the occurrence of huge coefficients, that one is prevented from obtaining the appropriate dimension and kernel information. Even with a good choice for a basis, during the intermediate computation the integers can be as large as 10^6. The choice of basis also has a marked effect on the density of the final system of equations, where the density is the percentage of nonzero coefficients occurring in the set of relations which express dependent monomials solely in terms of basis monomials. In the space of type (3,3,2), for example, the density varies from 43 to 83 per cent. Fortunately, density and coefficient size are highly correlated. We therefore choose bases with both smaller density and smaller maximum coefficient. There is an additional arithmetic property which governs our choice and is discussed in the next section in the first conjecture. These arithmetic properties clearly affect solvability but do not in general shed light on the structure of the Jordan ring R. There are, however, certain properties which are very exciting from the viewpoint of structure.

The first property relevant to the structure of R can be found with certain choices of basis. There exists a set of bases

B(i,j,k) for the subspaces of R through degree 9 such that all of the dependency relations are of the form

$$um = u_1b_1+u_2b_2+...+u_nb_n,$$

where u is a power of 2 and the b_p with p = 1,2,...,n are members of some B(i,j,k). Thus, for any (i,j,k) with i+j+k $\leq$ 9, there exists an r$\times$(r+d) matrix of linear equations whose leftmost r$\times$r submatrix is a diagonal matrix with powers of 2 on the diagonal. In such a matrix, d is the dimension of the subspace in question and r the number of dependent monomials -- the number of monomials outside the basis which remain after canonicalization with respect to both commutativity and distributivity.

The existence of these power of 2 systems establishes an important result for the free Jordan ring on 3 letters over a field of characteristics p $\neq$ 2. Recall that, although the information was obtained by restricting the computation to the integers, the discussion to this point has been over the rationals as coefficients. Although the entire effort, which rests upon the use of closing functions, is for characteristic 0, we now know that the subspaces of degree less than or equal to 9 of the free Jordan ring on 3 letters over a field of characteristic p $\neq$ 2 have the same dimension as those over a field of characteristic 0.

For the next property of interest, we turn to the contrasting behavior of the subspaces of R of even and odd degree. For each subspace of odd degree less than or equal to 9 which is isomorphic to its correspondent in the special Jordan ring, there exists a basis consisting of individual monomials m = m_fm_s such that all m_s are of degree 1. There are certain subspaces of even degree which also have such a basis, for example, those of type (5,1,0) and (7,1,0). On the other hand, no such basis exists for the space of type (2,2,0) even though there are precisely four monomials with m_s of degree 1 therein, and V(2,2,0) has dimension 4. Linear dependence is, of course, the obstacle. In fact many of the even degree subspaces do not have such a basis, among

which appear to be those of type (4,2,0), (3,3,0), (2,2,2), and
(4,2,2). (An exhaustive analysis has not yet been conducted for
these spaces.) The extension of the basis property to odd degree
subspaces not isomorphically mapped to the special Jordan ring
also fails. There are simply not enough individual monomials m
with m_s of degree 1 to match the dimension of either V(4,3,2) or
V(3,3,3), where m has the appropriate degree type.

   We next come to the property of R which relates the structure problem to the study of closing functions. For this
discussion, recall the fact that the set T(i,j,k) consists of
those permutations on 3 letters which preserve degree type.
Thus, T(3,3,2) consists of the identity and the transposition ab,
T(4,3,2) of the identity alone, and T(3,3,3) of all of S_3. Then
recall that a basis B(i,j,k) of individual monomials is closed
when, for all t in T(i,j,k), m is a member of B(i,j,k) if and
only if m' is, where m' and t(m) can be shown to be equal by
simply applying commutativity of multiplication. The property of
R under discussion states that, for every subspace of R of degree
less than or equal to 9, there exists a closed basis. In terms
of closing functions, we have the stronger statement that there
exists a function g for which each B(i,j,k)(g) with i+j+k $\leq$ 9 is
a closed basis of R(i,j,k)(g), where g is an initial segment of
a closing function. Although the function g is not itself a
closing function in that it is not defined for all (i,j,k), g
possesses the key properties of automating canonicalization of
both commutativity and distributivity, well-ordering the appropriate set of monomials, and mapping that set into the positive
integers. There are many choices for such mappings g into the
positive integers, each of which yields, in view of the results
of Section 4, a set of closed bases. Among the choices, there
are a number which are consistent with the previous two properties. There are, for example, functions g which are initial
segments through degree 9 of closing functions such that

1)   the $B(i,j,k)(g)$ with $i+j+k \leq 9$ are closed;

2)   where the mapping from the free ring to the special Jordan ring is an isomorphism, for all $m = m_f m_s$ in all $B(i,j,k)(g)$ with $i+j+k$ odd, m_s has degree 1; and

3)   the coefficient of the dependent monomial is a power of 2 for every relation expressing dependency in $R(i,j,k)(g)$ in terms solely of members of $B(i,j,k)(g)$.

This composite result naturally leads to a number of questions and conjectures, which brings us to the final section of the paper.

6.2. Conjectures and Open Questions

Proofs of or answers to the following should contribute to the solution of the puzzle of the general structure of the free Jordan ring R on 3 letters. Each of the following is a natural item for consideration, especially since the unnatural or inverted approach has brought us this far. Recall that the inverted approach rests heavily on the use of closing functions which in turn forces bases to be closed.

1. For all (i,j,k), we conjecture the existence of a set of bases $B(i,j,k)$ consisting of individual monomials such that when the coefficients are restricted to the integers, all of the corresponding dependency relations for monomials which are canonicalized for both commutativity of multiplication and for distributivity are power of 2 relations. In other words, all relations are of the form $pm = ...$, where p is a power of 2, m represents its class of commutative variants, m is recursively a product of basis monomials from the $B(i,j,k)$ in question, m is not a member of any of the $B(i,j,k)$, and the monomials to the right of the equality symbol are all members of a given $B(i,j,k)$ for that relation.

1'. We conjecture that the dimension of each subspace of R consisting of finite sums of monomials of degree type (i,j,k), for a given (i,j,k), with rational coefficients is unchanged by the replacement of the field of rationals by a field of characteristic $p \neq 2$.

2. For all (i,j,k), we conjecture that there exists a set of bases $B(i,j,k)$ such that each $B(i,j,k)$ is closed with respect to the subgroup of permutations on 3 letters consisting of those t with $t((i,j,k)) = (i,j,k)$, those which are degree type preserving.

3. For all (i,j,k), we conjecture that there exists a closing function -- a function g which satisfies properties p_0 to p_7 of Section 4. In fact we suspect that any of the initial segments g_1 which were employed during the investigation of R can be extended through any given (i,j,k) in a manner consistent with the demands of closing functions. If this is true, the inverted approach with the corresponding data structure of Section 2 can be used to study any given subspace of R.

4. Where $i \geq 2$, are all subspaces of type $(i,2,2)$ isomorphic to their correspondent in the special Jordan ring? We know from Table 2 that those of type $(2,2,2)$, $(3,2,2)$, $(4,2,2)$, and $(5,2,2)$ are mapped isomorphically. The dimension of $V(5,2,2)$ was previously unknown.

5. For all subspaces of R which are of odd degree and which map isomorphically to the special Jordan ring, we conjecture that the set of canonicalized monomials m therein with m_s of degree 1 forms a basis. Here, as expected, we mean canonicalized for both commutativity and for distributivity. In effect we are referring to the monomials m in some $R(i,j,k)(g)$ such that the degree of m_s equals 1. An even stronger conjecture appears to be true, namely, if the bases $B(i,j,k)(g)$ have been chosen for all subspaces up to and including degree d_1 for some initial segment g

of a closing function, and if d_1 is an even integer, then for
each R(i',j',k')(g) with i'+j'+k' = d_1+1 which maps isomorphically to the special Jordan ring one obtains a basis by simply multiplying on the right by a or by b or by c those monomials which
are both basis monomials relative to g and which are of degree
d_1. For example, if R(i',j',0)(g) is such a subspace, then one
multiplies by a the members of B(i'-1,j',0)(g), and by b the
members of B(i',j'-1,0)(g), and obtains a basis for R(i',j',0)(g).

We come now to the last conjecture which, as with 3 above,
relates the study of closing functions to the structure problem
for R. Note that, by applying the results of Section 4, a proof
of the third conjecture would yield a proof of the second. In
particular, Corollary 1 of Section 4 states that all bases
B(i,j,k)(g) relative to a closing function g are closed. Closed
bases appear to force a connection of a quite special type between R and the special Jordan ring. This connection is given in
6 below. In this regard, first recall that a monomial m is fixed
if and only if there exists a transposition t such that t(m) and
m can be shown to be equal with the application of commutativity.
(An equivalent form of this concept is found in Section 3.1.)
This definition can be extended in a natural way to the symmetric
elements of the special Jordan ring. Some examples of symmetric
elements are, abc+cba, acb+bca, aabb+bbaa, and aabbab+babbaa.
The monomial (bb)(aa) is fixed, and it is fixed under ab. The
first three symmetric elements are also fixed, but the last is
not. If one applies to each of the first three respectively the
transpositions ac, ab, and ab, one simply obtains the same three
elements if commutativity of addition is applied. For the fourth
symmetric element, one applies ab, and obtains after canonicalization for commutativity of addition the symmetric element
abaabb+bbaaba. Since ab is the only possible transposition which
can possibly fix the fourth symmetric element, that symmetric
element is said to be paired. Similarly, the monomial ((aa)b)b is
a paired monomial. Finally, we recall the fact that the symmetric

elements of a given degree type are a basis for the corresponding subspace of the special Jordan ring, which brings us to the conjecture.

6. If B(i,j,k) is a closed basis consisting of individual monomials for the subspace V(i,j,k) of R, and if V(i,j,k) is isomorphic to the corresponding subspace of the special Jordan ring, then we conjecture that the number of fixed and paired monomials in B(i,j,k) must be the same as that in the basis of symmetric elements for the correspondent of V(i,j,k). For clarity here, consider the subspace for the symmetric elements of type (2,2,2). The symmetric element aabbcc+ccbbaa is fixed under ac and not under ab or bc. It is, however, classed as fixed. The paired elements are those which are not fixed under any transposition. The symmetric elements of type (2,2,2) are partitioned into the fixed and paired classes with 18 fixed and 30 paired. The closed bases which have been studied for R(2,2,2)(g), and hence for V(2,2,2), are such that each consists of 18 fixed and 30 paired monomials.

In conclusion, since there exist many functions g through degree 9 which behave like closing functions and which, therefore, yield closed bases, it appears that there exist functions which can be extended through degree d_1 in a manner consistent with properties p_0 to p_7 for any arbitrarily chosen positive integer d_1. Further, it appears that, among the sets of closed bases, there exist many which simultaneously satisfy where meaningful the conjectures above. For example, there appear to be many which match the fixed and paired count property of 6 and at the same time satisfy the power of 2 property of the first conjecture and the requirement for bases of the appropriate subset of the set of odd degree subspaces of 5.

Acknowledgments

We wish to thank Dr. J. Marshall Osborn of the Mathematics Department of the University of Wisconsin for his suggestion of studying the free Jordan ring on 3 letters with the aid of a computer. We wish to thank him also for his assistance with certain technical details of the paper. We then wish to thank in chronological order the following individuals for various contributions. Mr. William J. Snow programmed the first version of the approach taken herein, and it was partly at his instigation that symbolic representation was replaced by integral representation. Dr. Joseph H. Mayne provided us with a useful formula for evaluating determinants related to the regular representation of S_3. Mr. Michael Tubin programmed a number of the subroutines contained in the present system and in general contributed to program efficiency. Mr. Robert Veroff designed and implemented the programs which were instrumental in obtaining the results for the kernel. Mr. Joseph Glover contributed to the technique for choosing uniformly structured bases. We wish also to thank Miss Judy Beumer for the preparation of this lengthy manuscript.

This work was performed under the auspices of the U.S. Energy Research and Development Administration.

References

[1] C. M. Glennie, *Some identities valid in special Jordan algebras but not valid in all Jordan algebras*, Pacific Journal of Mathematics, Volume 16, Number 1, 1966.

[2] N. Jacobson, *Structure and representations of Jordan algebras*, American Mathematical Society Colloquium Publications, Volume 39, 1968, pp. 40-49.

[3] D. E. Knuth, *The art of computer programming, volume 3, sorting and searching*, Addison-Wesley, 1973.

Applied Mathematics Division
Argonne National Laboratory
Argonne, Illinois 60439

ON THE INVARIANTS OF A LIE GROUP. I

Hans Zassenhaus

1.   In 1932 Casimir [2] in response to a question of B. L. van
der Waerden produced a nontrivial second degree expression in
terms of the infinitesimal transformation algebra $L = L(G)$ of
a simply connected semisimple Lie group G commuting elementwise with G for all continuous representations of finite
degree. In terms of an R-basis $X_1, X_2, \ldots, X_n$ of L with Lie
multiplication rule

(1a) $[X_i, X_k] = \Sigma_{j=1}^n \gamma_{ik}{}^j x_j$ $(\gamma_{ik}{}^j \in R;\ i, k, j = 1, 2, \ldots, n)$

of the Lie algebra $L(G)$ of G over the real number field R
his expression was of the form

(1b) $C_A = \Sigma_{i,k=1}^n \alpha^{ik} X_i X_k$ $(A = (\alpha^{ik}) = A^T \in R^{n \times n}).$

For every Banach space M providing a continuous representation

(1c) $G \to \mathrm{End}_R M:\ \Psi$

of G by bounded operators of M we have the linear representation

(1d) $L \to \mathrm{End}_R M:\ \Delta$

(1e) $X \Delta = \lim_{\alpha \to 0} \alpha^{-1} (\exp(\alpha X)^\Psi - 1_M)$ $(X \in L)$

of L; it is demanded that the expression

(1f) $C_A \Delta = \Sigma_{i,k=1}^n \alpha^{ik} (X_i \Delta)(X_k \Delta)$

commute elementwise with G^{Ψ}:

(1g) $$[C_A\Delta, g^{\Psi}] = 0 \qquad (g \in G)$$

implying $0 = [C_A\Delta, \exp(\alpha X)] = [C_A\Delta, \alpha^{-1}(\exp(\alpha X)^{\Psi} - 1_M)]$ if $0 \neq \alpha \in R$, $X \in L$ and therefore according to (1e)

(1h) $$0 = [C_A\Delta, X\Delta] \qquad (X \in L).$$

Conversely, if for any representation (1d) of L over R by bounded operators of the Banach space M the expression (1f) satisfies the conditions (1h) then there is precisely one continuous representation Ψ of G by bounded operators of M which satisfies (1e) and G is generated by the subset

$$\exp L = \{\exp X | X \in L\}$$

of G such that

(1i) $$(\exp X)^{\Psi} = \exp(X\Delta) \qquad (X \in L).$$

It follows that $C_A\Delta$ commutes elementwise with G^{Ψ}.

Thus it is seen that the conditions (1h) on C_A in terms of the infinitesimal ring which were given by Casimir already provide the equivalent to the conditions posed by van der Waerden.

2.   In modern terminology the *Casimir operator* (1a) is interpreted as an element of the universal embedding àlgebra [1,3,4] $U(L) = U_R(L)$, of L over R. Observing that any representation (1d) of L by bounded operators of the Banach space M gives rise to precisely one representation $\Delta^{U(L)}$ of the associative algebra $U(L)$ by bounded operators of M restricting to Δ on L and that the intersection of the kernels of the $\Delta^{U(L)}$'s is 0 it follows that C_A meets the condition (1h) for all Δ's precisely if

(2a) $$[C_A, X_h] = 0 \qquad (1 \leq h \leq n).$$

or, equivalently,

(2b) $\sum_{h=1}^{n} \alpha^{ih} \gamma_{hj}^{\ k} = \sum_{h=1}^{n} \alpha^{kh} \gamma_{jh}^{\ i}$ $(i,j,k = 1,2,\ldots,n)$.

One may interpret G as the group generated by exp L where the exponential series

$$\exp X = \sum_{j=0}^{\infty} n!^{-1} X^n \qquad (X \in L)$$

are formed in a Banach completion of the associative algebra $U(L)$ for some suitable norm. There always are norms of $U(L)$.

As a matter of fact, any neighborhood N of 0 in L defines the *germ*

$$\exp N = \{X \,|\, X \in \exp N\}$$

of G such that G is generated by exp X with the relators

$$g_1 g_2 = g \qquad (g_1, g_2, g \in \exp N)$$

that are valid already in exp N.

This group is indeed continuous and simply connected for the Banach topologization.

In order to evaluate (1h) we follow Whitehead's [4] procedure of using the nondegenerate Cartan-Killing form with coefficients

(2c) $g_{ik} = \text{tr}((X_i \text{ad})(X_k \text{ad})) = \sum_{h,j=1}^{n} \gamma_{hi}^{\ j} \gamma_{jk}^{\ k}$

$(i,k = 1,2,\ldots,n)$

as fundamental tensor to raise and lower indices. Then (1h) is equivalent to the symmetry condition

(2d) $\gamma_{ijk} = \gamma_{kij}$ $(i,j,k = 1,2,\ldots,n)$

for the multiplication tensor which together with the anti-symmetry of Lie multiplication amounts to the $L(G)$-admissibility condition

(2e) $([a,b] \times c) f_A = (a \times [b,c]) f_A$

for the symmetric bilinear form

(2f) $(\Sigma_{i=1}^{n} \xi^i X_i \times \Sigma_{k=1}^{n} \eta^k X_k) f_A = \Sigma_{i,k=1}^{n} \xi^i \eta^k \alpha_{ik}$

defined on $L(G)$ for the covariant matrix (α_{ik}). Any trace
bilinear form of a finite degree representation of $L(G)$ is
symmetric and $L(G)$-admissible, in particular the choice

(2g) $(\alpha_{ik}) = (g_{ik})$ or $(\alpha^{ik}) = (g_{ik})^{-1}$

provides us with the Cartan-Killing bilinear form.

Its nondegeneracy for semisimple Lie algebras results in
the special Casimir operator

(2h) $C = \Sigma_{i=1}^{n} X_i X^i$

which is actually the one given explicitly by Casimir. Of
course all this can equally well be done for any field of reference F of zero characteristic.

The mapping of f_A on C_A provides a linear monomorphism of
the linear space $B(L)$ of the $L(G)$-admissible symmetric bilinear
forms on L into $U(L)$ the image space of which is precisely
the intersection of the center of $U(L)$ with L^2. The F-dim-
ension of $B(L)$ equals the number of absolutely simple components of L.

3.   Casimir's construction is the first step towards the study
of the invariant polynomials, the rational invariants and the
invariants of a Lie group which will form the object of this
paper.

Let L be a Lie algebra over a field F of characteristic
0, let $U(L) = U_F(L)$ be the universal embedding algebra of L
over F.

Definition 1.   The elements of $U(L)$ commuting by ordinary
multiplication with a given generator set of L over F are
said to be the *invariant polynomials* or *Casimir polynomials*
over L.

The invariant polynomials over L form a commutative and
associative F-algebra $CU(L)$ without divisors of zero, the
center of $U(L)$:

$$CU(L) = \{c \,|\, c \in U(L) \text{ and for all } u \in U(L), cu = uc\}$$
$$= \{c \,|\, c \in U(L) \text{ and } [c, U(L)] = 0\}.$$

There is the well known filtration of $U(L)$ by means of the
associative power series

(3a) $0 \subset L \subset L + L^2 \subset L + L^2 + L^3 \subset \ldots \subset U_F(L)$

of L which is used to define recursively a generator set
$\{c_{ij} \,|\, j = 1, 2, \ldots, \zeta(i); \; i = 1, 2, \ldots\}$ of $CU(L)$ over F as follows:

1)   Let $\{c_{1j} \,|\, j = 1, 2, \ldots, \zeta(1)\}$ be any F-basis of the center of L.

2)   If $i > 1$, then let

$$\{c_{ij} \,|\, j = 1, 2, \ldots, \zeta(i); \; i = 1, 2, \ldots\}$$

be any F-basis of

(3b) $C^{(i)} = CU(L) \cap (L + L^2 + \ldots + L^i)$

modulo the F-linear subspace

(3c) $C^{(i)*} = \Sigma_{j=1}^{i-1} \, C^{(j)} C^{(i-j)}$

generated by the products

$$c_{i_1 j_1} \; c_{i_2 j_2} \; \cdots \; c_{i_r j_r}$$

$$(1 \leq i_1 \leq i_2 \leq \ldots \leq i_r, \; \sum_{h=1}^{r} i_h \leq i, \; r > 1$$

$$\text{and } 1 \leq j_h \leq \zeta(i_h), \; h = 1, 2, \ldots, r)$$

The *filtration generator sets* of $CU(L)$ described above are 'essentially' uniquely determined. One is obtained from the other by modification of c_{ij} modulo $C^{(i)*}$ and by application of F-linear automorphisms of $C^{(i)}/C^{(i)*}$. Thus, we obtain the invariants

(3d) $\zeta(i,L) = \zeta(i)$ $(i = 1, 2, \ldots)$

indicating the number of elements of 'dimension i' belonging to a filtration generator set of $CU(L)$.

Definition 2.   The *Casimir rank* of L is defined as the degree of transcendency $cr(L)$, of $CU(L)$ over F. Surely we have the inequality

(3e) $\sum_{i=1}^{\infty} \zeta(i,L) \geq cr(L)$.

Definition 3.   The *Casimir dimension* $\delta(L)$ of L is defined as infinity in case there are infinitely many positive $\delta(i,L)$. It is defined as the maximum value of i for which $\delta(i,L) > 0$ in case there are only finitely many $\delta(i,L) > 0$.

4.   The construction is simplified by introducing a *grading* of the F-linear space U(L) that is compatible with the given filtration according to

(4a) $(L + L^2 + \ldots + L^{i-1}) + U(L)_i = L + L^2 + \ldots + L^i$

                                $(i = 1, 2, \ldots)$

and which satisfies the L-invariance condition

(4b) $$[U(L)_i, L] \subseteq U(L)_i.$$

The ith component, $U(L)_i$, is defined as the module generated by the *symmetrized sums*

(4c) $$(a_1 a_2 \cdots a_i)\sigma_i = \frac{1}{i!} \sum_{\pi \in S_i} a_{1\pi} a_{2\pi} \cdots a_{i\pi}$$

for any i elements $a_1, a_2, \ldots, a_i$ of L. Evidently $U(L)_i$ is an F-linear subspace of $U(L)$. Its L-invariance follows from the rule

(4d) $$[(a_1 a_2 \cdots a_i)\sigma_i, x] = \frac{1}{i!} \sum_{\pi \in S_i} [a_{1\pi} a_{2\pi} \cdots a_{i\pi}, x]$$

$$= \frac{1}{i!} \sum_{\pi \in S_i} ([a_{1\pi}, x]a_{2\pi} \cdots a_{i\pi} + \ldots + a_{1\pi} \cdots a_{(i-1)\pi}[a_{i\pi}, x])$$

$$= ([a_1, x]a_2 \cdots a_i)\sigma_i + \ldots + (a_1 \cdots a_{i-1}[a_i, x])\sigma_i.$$

In order to verify (3a) and to extend σ_i to an endomorphism of L^i over F we use the natural epimorphism

(4e) $$F\langle L \rangle \to U(L): \quad \varepsilon$$
$$a\varepsilon = a \quad (a \in L)$$

of the free associative F-algebra $F\langle L \rangle$ generated by the F-linear space L on the universal embedding algebra $U(L)$ over F. Furthermore, we use the multiplicative grading

(4f) $$F\langle L \rangle = \sum_{i=1}^{\infty} F\langle L \rangle_i$$

where $F\langle L \rangle_i$ is the linear F-subspace of $F\langle L \rangle$ generated by the products of any i elements of L such that

(4g) $$F\langle L \rangle_i \cong \underbrace{L \times_F L \times_F \ldots \times_F L}_{i \text{ times}},$$

(4h) $$F\langle L\rangle_i F\langle L\rangle_k = F\langle L\rangle_{i+k}.$$

The symmetrization mapping τ_i that maps the generator $a_1 a_2 \cdots a_i$ of $F\langle L\rangle_i$ on

(4i) $$(a_1 a_2 \cdots a_i)\tau_i = \frac{1}{i!} \sum_{\pi \in S_i} a_{1\pi} a_{2\pi} \cdots a_{i\pi}$$

preserves the defining relations of the F-linear space $F\langle L\rangle_i$ so that it can be extended uniquely to an F-linear endomorphism of $F\langle L\rangle_i$ (S_i the symmetric permutation group of $\{1,2,\ldots,i\}$).

This endomorphism τ_i is idempotent and we have

(4j) $$\tau_i \varepsilon = (F\langle L\rangle_i \mid \varepsilon)\sigma_i, \quad F\langle L\rangle_i \tau_i \varepsilon = U(L)_i.$$

Finally, we use the natural F-homomorphism

(4k) $$F\langle L\rangle \to F[L]: \eta,$$
                         $$a\eta = a \quad (a \in L)$$

of the free associative F-algebra $F\langle L\rangle$ over L into the free unital associative and commutative algebra $F[L]$ (polynomial algebra over F) leading to the grading

(4l) $$F[L] = \sum_{i=0}^{\infty} F[L]_i$$

by the homogeneous polynomials over L such that

(4m) $$F[L]_0 = F$$
                         $$F[L]_i = F\langle L\rangle_i \eta \quad (i > 0).$$

According to the Birkhoff-Witt-Poincaré [1,3,5] embedding construction there is the F-epimorphism

(4n) $$L + L^2 + \ldots + L^i \to F[L]_i: \pi_i$$
                $$(a_1 a_2 \cdots a_j)\pi_i = 0 \quad (1 \le j < i)$$
                $$(a_1 a_2 \cdots a_i)\pi_i = a_1 a_2 \cdots a_i.$$

On the other hand the kernel of the F-epimorphism of $F\langle L\rangle_i$ on $F[L]_i$ obtained by restriction of η coincides with the kernel of τ_i so that because of the idempotency of σ_1

(40)
$$F\langle L\rangle_i \;\big|\; \eta = \tau_i \eta$$

and the restriction of $\tau_i \eta$ to $F\langle L\rangle_i \tau_i$ is an F-isomorphism of $F\langle L\rangle_i \tau_i$ on $F[L]_i$.

It follows that there is the F-endomorphism

(4p)
$$L^i \;\rightarrow\; U(L)_i : \;\; \sigma_i$$

$$u\varepsilon\sigma_i = u\tau_i \quad (u \in F\langle L\rangle_i)$$

such that π_i restricts on $U(L)_i$ to an F-isomorphism of $U(L)_i$ on $F[L]_i$. It also follows that

(4q)
$$U(L)_i \cap (L + L^2 + \ldots + L^{i-1}) = 0$$

so that the grading condition (4a) applies.

Beside the L-invariance (4b) the grading also satisfies the further multiplicative condition

(4r)
$$(uv)\sigma_{i+k} = (u\sigma_i \; v\sigma_k)\sigma_{i+k}$$
$$(u \in L + L^2 + \ldots + L^i, v \in L + L^2 + \ldots + L^k).$$

It is convenient to extend the definition of the symmetrization operators σ_i to $U(L)$ by setting

(4s)
$$(\Sigma_{j=1}^{t} u_j)\sigma_i = u_i \qquad (u_j \in U(L)_j)$$

In this way they appear as the system of orthogonal decomposition operators pertaining to the direct module theoretic decomposition

(4t)
$$U(L) = \Sigma_{i=1}^{\infty} U(L)_i$$

commuting with the Lie action of L on $U(L)$.

5.   Using the grading of $U(L)$ the previous construction of a filtration generator set of $CU(L)$ is simplified to the construc-

tion of a *graded generator set* of CU(L) by demanding that each generator belongs to one of the grading components. This condition amounts to the choice of the c_{ij} $(j = 1,2,\ldots,\zeta(i))$ as an F-basis of

(5a) $$CU(L)_i = CU(L) \cap U(L)_i = C^{(i)}\sigma_i$$

modulo the F-linear subspace

(5b) $$CU(L)_i^* = (\Sigma_{j=1}^{i-1} CU(L)_j CU(L)_{i-j})\sigma_i$$

$$= C^{(i)*}\sigma_i.$$

Observe that because of the L-invariance of $U(L)_i$ any element of U(L) belongs to the center precisely if all its grading components belong to the center. In other words we have the direct decomposition

(5c) $$CU(L) = \Sigma_{i=1}^{\infty} CU(L)_i$$

of the center of U(L) compatible with the grading (3a).

6.   Choosing a well ordered F-basis $X_1, X_2, \ldots$ of L with multiplication constants defined by

(6a) $[X_i, X_k] = \Sigma_{j \geq 1} \gamma_{ik}^j X_j$ $(\gamma_{ik}^j \in F; \quad i \geq 1, \ k \geq 1, \ j \geq 1)$

we observe that F[L] is simply the polynomial ring in the independent variables $X_1, X_2, \ldots$ The equation (4d) is equivalent to the rule

(6b) $$[(a_1 a_2 \cdots a_i)\sigma_i, X]\pi_i = (a_1 a_2 \cdots a_i)\pi_i d_X$$

$$(a_1, a_2, \ldots, a_i \in L, \ X \in L)$$

where d_X is the F-derivation of the polynomial F-algebra F[L] over L defined by

(6c)
$$X = \Sigma_{i \geq 1} \zeta^i X_i \quad (\zeta^i \in F),$$

(6d)
$$Pd_{X_i} = \Sigma_{k \geq 1} \frac{\partial}{\partial X_k} (P) P_{ik} \quad (P \in F[L]),$$

(6e)
$$P_{ik} = \Sigma_{j=1}^{n} - \gamma_{ik}^{j} X_j,$$

(6f)
$$d_X = \Sigma_{i \geq 1} \zeta^i d_{X_i}.$$

For this interpretation the task of determining $CU(L)$ is equivalent to establishing the polynomial solutions of the system of partial differential equations

(6g)
$$Pd_{X_i} = 0 \qquad (i \geq 1, P \in F[L]).$$

7.   We observe that the mapping

(7a)
$$L \to End_F F[L]: \quad d$$
$$Xd = d_X$$

is a representation of L with $F[L]$ as representation space and with CL as kernel. Its restriction to L is equivalent to the adjoint representation of L. The F-derivations of $F[L]$ form a Lie algebra $Der_F F[L]$ which is a right $F[L]$-module with $F[L]$-basis formed by the partial differentiation $\frac{\partial}{\partial X_i}$.

Definition 4.   If

(7b)
$$dim_F L = n$$

is finite then the $F[L]$-*rank* of the F-Lie algebra Ld is a non-negative integer $\rho_F(L) = \rho(L)$ which is not larger than n. This important invariant of the Lie algebra L is equal to the rank of the differential matrix (P_{ik}) corresponding to the F-basis $X_1, X_2, \ldots, X_n$ of L.

8.   It is well known that U(L) has no divisors of zero and
that any two elements of U(L) have both common right multipla
and common left multipla. As a consequence the Ore quotient
ring of U(L) is a division algebra.

$$D(L) = (U(L)\backslash 0)^{-1}U(L)$$
$$= U(L)(U(L)\backslash 0)^{-1}$$

over F. Its center, CD(L), is a field extension of F containing
the center of U(L) and its quotient field is QCU(L) though it is
not necessarily equal to it.

Definition 5.   The elements of CD(L) are called the *rational
invariants of L*. The transcendency degree over F of the field
CD(L) formed by the rational invariants of L is not smaller
than the Casimir rank of L over F. In order to study the re-
lationship between the rational invariants and the invariant
polynomials we form the quotient field Q(CU(L)) of CU(L) and the
central quotient ring

(8a)   R(L) = Q(CU(L)) + Q(CU(L)) U(L)

of U(L). There is a Q(CU(L))-basis $1, Y_1, \ldots, Y_\nu$ of

(8b)   Q(L) = Q(CU(L)) + Q(CU(L))L

with $Y_1, \ldots, Y_\nu$ in L. Using the Birkhoff straightening method
we find the Q(CU(L))-Birkhoff basis

(8c) $Y_1^{\alpha_1} Y_2^{\alpha_2} \ldots Y_\nu^{\alpha_\nu}$ $(0 \le \alpha_i, \ 1 \le i \le \nu)$

of Q(CU(L))U(L). Hence, there is the Q(CU(L))-basis

(8d) $Y_1^{\alpha_1} \ldots Y_\nu^{\alpha_\nu}$ $(\Sigma_{i=1}^{\nu} \alpha_i = \alpha, \ 0 \le \alpha_i, \ 1 \le i \le \nu)$

of $Q(L)^\alpha$ modulo $Q(L)^{\alpha-1}$ $(\alpha > 0)$ and there is the Q(CU(L))-
epimorphism

(8e) $Q(L)^\alpha \to Q(CU(L))[Y_1,\ldots,Y_\nu]_\alpha: \quad \varepsilon_\alpha$

$$(Y_1^{\alpha_1}\ldots Y_\nu^{\alpha_\nu})\varepsilon_\alpha = \begin{vmatrix} Y_1^{\alpha_1}\ldots Y_\nu^{\alpha_\nu} & \text{if } \Sigma_{i=1}^\nu \alpha_i = \alpha \\ \\ 0 & \text{if } \Sigma_{i=1}^\nu \alpha_i < \alpha \end{vmatrix}$$

$$(0 \le \alpha_i, \ 1 \le i \le \nu, \ \Sigma_{i=1}^\nu \alpha_i \le \alpha)$$

of $Q(L)^\alpha$ on the $Q(CU(L))$-module of the homogeneous polynomials of degree α in $Y_1,\ldots Y_\nu$ over $Q(CU(L))$. The kernel of ε_α is $Q(L)^{\alpha-1} (\alpha \ge 1)$. The epimorphisms ε_α have the multiplicative properties:

(8f) $$(xy)\varepsilon_{\alpha+\beta} = (x\varepsilon_\alpha)(y\varepsilon_\beta)$$

$$(x \in Q(L)^\alpha, \quad y \in Q(L)^\beta, \quad (0 \le \alpha, \ 0 \le \beta)$$

(8g) $$[x,u]\varepsilon_\alpha = (x\varepsilon_\alpha)d_u \quad (u \in Q(L)).$$

Suppose there is the rational invariant

$$x = ab^{-1}$$

for which

$$a \in Q(L)^\alpha, \ a \notin Q(L)^{\alpha-1}, \ b \in Q(L)^\beta, \ b \notin Q(L)^{\beta-1}, \ \beta > 0.$$

Since x commutes with all elements of $D(L)$ it also commutes with b. But since b itself commutes with b it follows that

(8h) $$[a,b] = 0.$$

Furthermore for all u of L

$$0 = [x,u] = [ab^{-1},u]$$

$$= [a,u]b^{-1} - ab^{-1}[b,u]b^{-1}$$

hence

(8i) $$[a,u]b = a[b,u],$$

hence for all u of $Q(L)$

(8j) $[a,u]\varepsilon_\alpha b\varepsilon_\beta = a\varepsilon_\alpha [b,u]\varepsilon_\beta.$

There are non-zero homogeneous polynomials p, q, c, of
$Y_1, Y_2, \ldots, Y_\nu$ over $Q(CU(L))$ such that

(8k) $a\varepsilon_\alpha = pc, \quad b\varepsilon_\beta = qc$

and p,q are mutually prime. Now (8j) implies that

$$((pd_u)c + p(cd_u))qc = pc((qd_u)c + q(cd_u),$$

$$(pd_u)q = p(qd_u).$$

Since p,q have no common divisor it follows that

$$p \mid (pd_u), \qquad q \mid (qd_u)$$

and since the degrees of p and pd_u, q and qd_u are equal (or
else $pd_u = 0$ or $qd_u = 0$) it follows that there is an element
$u\lambda$ of $Q(CU(L))$ for which

(8l) $pd_u = u\lambda p, \quad qd_u = u\lambda$ $(u \in Q(L)).$

The mapping

(8m) $Q(L) \rightarrow Q(CU(L)): \quad \lambda$

is a $Q(CU(L))$-homomorphism of the $Q(CU(L))$-Lie algebra $Q(L)$ in
$Q(CU(L))$ mapping $Q(CU(L))$ on 0.

Theorem 1.   If the Lie algebra L over the field F is perfect or
nilpotent then the rational invariants of L form the quotient
field of the center of $U(L)$.

Proof.   The assumption that L is perfect, i.e., $[L,L] = L$, or
that L is nilpotent implies that the mapping λ defined by (8m),
(8n) always is 0. This means that p and q both are in $Q(CU(L))$
and that $\alpha = \beta$.

Hence also $x - pq^{-1} = (a - pq^{-1}b)b^{-1}$ is a rational invariant.
But since

$$a - pq^{-1}b \in Q(L)^{\alpha-1} = Q(L)^{\beta-1}$$

it follows by the same argument that

$$a - pq^{-1}b = 0,$$

hence

$$x = ab^{-1} = pq^{-1} \in Q(C(L)).$$

Note that it can very well occur for non-perfect and non-nilpotent Lie algebras that there are rational invariants which are not quotients of invariant polynomials. For example, the solvable, non-nilpotent Lie algebra L with multiplication table

[ ]	X_1	X_2	X_3
X_1	0	0	X_1
X_2	0	0	X_2
X_3	$-X_1$	$-X_2$	0

over a field F of zero characteristic has no polynomial invariant $\neq 0$. But there is the rational invariant $X_1 X_2^{-1}$.

9.   Every F-derivation d of the Lie algebra L extends uniquely to an F-derivation $d^{U(L)}$ of U(L) according to the rule

(9a) $(a_1 a_2 \cdots a_i)d^{U(L)} = (a_1 d)a_2 \cdots a_i + \ldots + a_1 \cdots a_{i-1}(a_i d)$

$$(a_1, a_2, \ldots, a_i \in L).$$

Every derivation d of U(L) extends uniquely to a derivation $d^{D(L)}$ of D(L) according to the rule

(9b) $(ab^{-1})d = (ad)b^{-1} - ab^{-1}(bd)b^{-1}$

Any F-derivation d of L extends uniquely to the F-derivation

(9c) $$d^{D(L)} = (d^{((L)})^{D(L)}$$

of D(L).

There is the canonical mapping

9(d) $\qquad \text{Nor}_{D(L)}(L) \rightarrow \text{Der}_F(L): \quad \lambda$

$$a(X\lambda) = [a,X] \qquad (a \in L, \; x \in \text{Nor}_{D(L)}(L))$$

of the normalizer

(9e) $\qquad \text{Nor}_{D(L)}(L) = \{x \mid x \in D(L) \text{ and } [L,x] \subseteq L\}$

of L in D(L) onto the Lie algebra $\text{Der}_F(L)$ formed by the F-derivations of F. What is the image of λ? What is the transcendency degree of U(L) over F? What is the transcendency degree of CD(L) over F?

10. In terms of the preceding definitions the following answers were found:

Theorem 2. a) For any Lie algebra L over a field F the λ-image of the Lie normalizer of L in the division algebra D(L) generated by L over F is an ideal of $\text{Der}_F(L)$ contained in the ideal $\text{Der}_{CD(L)}L$ of L formed by all F-derivations d of L annihilating C(L) according to

(10a) $\qquad d^{D(L)}(CD(L)) = 0.$

b) If L is of finite dimension over the field F of characteristic 0 then

(10b) $\qquad \text{Nor}_{D(L)}(L)\lambda = \text{Der}_{CD(L)}(L).$

Corollary. For every finite-dimensional F-ideal A of a Lie algebra L over a field F of zero characteristic the elements x of L for which $CD(A)(A \mid ad_L x)^{D(L)} = 0$ form an F-ideal $C_{CD(A)}(A)$ of L containing A.

Theorem 3. a) If L is of finite dimension over the field F of zero characteristic then any filtration generator set of CU(L) is finite.

b) CD(L) is purely transcendental of transcendency

degree over F not larger than $\dim_F(L) - \rho_F(L) = n - \rho(L)$.

      c) Equality obtains in b) precisely if L is an algebraic Lie algebra.

      d) The transcendency degree of CD(L) over F is equal to the Casimir rank of L over F precisely if CD(L) is the quotient field of CU(L).

References

[1] Garrett Birkhoff, *Representability of Lie algebras and Lie groups by matrices*, Annals. of Math. (2) 38(1937) 526-532.

[2] H. Casimir, *Bemerkung zur Theorie der Storungen in Hyperfeinstrukturen*, Zeit. f. Physik 77 (1932), 811-814.

[3] H. Poincaré, *Sur les groupes continus*, Trans. Cambridge Phil. Soc. 18 (1899), 220-255.

[4] L. Abellanas and L. Martinez Alonso, *A general setting for Casimir invariants*, J. of Math. Physics 16 (1975), 1580-1584.

[5] E. Witt, *Freie Darstellung Lieschen Ringe*, J. Reine Angew. Math. 177 (1937), 152-160.

Department of Mathematics
The Ohio State University
Columbus, Ohio   43210

COMPUTATION OF CASIMIR INVARIANTS
OF LIE ALGEBRAS

C. W. Conatser
and
P. L. Huddleston

1. Introduction

The Casimir invariants of a Lie algebra are of considerable
interest in mathematics and in theoretical physics. In the
representation theory of Lie groups and Lie algebras, the eigenvalues of the Casimir operators can be used to label the irreducible representations. In the theory of special functions, the
various special functions can be seen to arise as eigenfunctions
of Casimir operators. In physical applications, the Casimir
operators of the Lie algebra of a Lie symmetry group of a physical
system form part of a set of commuting operators whose eigenvalues
are the quantum numbers of the system.

The number and form of the Casimir invariants of the semisimple Lie algebras are known [9]. For years, the Casimir
invariants for non-semisimple Lie algebras were worked out by
hand on a case by case basis. Then it was observed [3,8] that
the invariants of any finite-dimensional Lie algebra could be
determined from the solutions of a system of first-order partial
differential equations. The invariants that can be determined
in this way include, besides the polynomial Casimir invariants,
the so-called rational and general invariants [1]. Recently,
this approach has been systematically applied to all real Lie
algebras of dimension less than or equal to five, to all real
nilpotent Lie algebras of dimension six [5], and to various
other Lie algebras of higher dimension [6,7].

In this paper, it is shown that the Casimir invariants of any degree of an arbitrary finite-dimensional Lie algebra can be determined from the solutions of a system of homogeneous linear equations depending on the structure constants of the algebra. In Section 2 the general form of the system of linear equations that must be satisfied by the coefficients of a homogeneous polynomial Casimir invariant of degree d is presented. This makes the calculation of Casimir invariants amenable to computation by digital computer. In Section 3 the systems of equations for Casimir invariants of degrees one, two, and three are written out explicitly. In Section 4 algorithms and their computer imple-mentation for generating and solving the systems of equations for Casimir invariants are described.

2.   The System of Equations For Casimir Invariants of Degree d

Let L be an arbitrary real Lie algebra of dimension n, with a basis denoted by X_1, X_2,...,X_n. Let the structure constants for L in this basis be c_{ijk}, then the Lie brackets or commutators of the basis elements are given by

$$[X_i, X_j] = \sum_{k=1}^{n} c_{ijk} X_k,$$

where i, j = 1, 2,...,n.

Let U(L) be the universal enveloping algebra of L, which can be thought of as the space of all polynomials in $X_1, X_2, \ldots, X_n$ with the Lie bracket [X,Y] of elements X and Y of L being iden-tified with the commutator XY-YX [2]. Now for each degree d and each choice of d indices $1 \leq i_k \leq n$, k=1,...,d, define

$$\phi_{i_1,\ldots,i_d} = \sum_{\sigma \in S_d} X_{i_{\sigma(1)}} X_{i_{\sigma(2)}} \ldots X_{i_{\sigma(d)}},$$

where S_d is the symmetric group on d objects. Observe that

$$\phi_{i_1,\ldots,i_d} = \phi_{i_{\sigma(1)},\ldots,i_{\sigma(d)}} \qquad \text{for all } \sigma\epsilon S_d. \quad \text{Those } \phi_{i_1,\ldots,i_d}$$

with ordered indices $i_1 \leq i_2 \leq \ldots \leq i_d$ are linearly independent and
so form a basis for their linear span, which we denote by $U^{(d)}(L)$
and call the space of homogeneous polynomials of degree d. Thus
$U(L)$ has the direct sum decomposition:

$$U(L) = \bigoplus_{d=0}^{\infty} U^{(d)}(L).$$

The reason for choosing the symmetrized basis instead of the
usual Poincare-Birkhoff-Witt basis is that each of the $U^{(d)}(L)$
is invariant under the adjoint action of L on $U(L)$ [4].

A Casimir invariant C of L is an element of $U(L)$ that has
vanishing Lie bracket with all basis elements of L:

$$(2.1) \qquad\qquad [C, X_k] = 0,$$

for all k=1,...,n. The set of all Casimir invariants forms the
center $ZU(L)$ of $U(L)$, and has a corresponding direct sum
structure

$$ZU(L) = \bigoplus_{d=0}^{\infty} ZU^{(d)}(L),$$

where each component is the space of homogeneous polynomial
invariants of degree d. This structure makes it very convenient to calculate the Casimir invariants one degree at a time.

The invariance of each $U^{(d)}(L)$ under the adjoint action of
L on $U(L)$ means that the Lie bracket $[\phi_{i_1,\ldots,i_d}, X_k]$ can be
written as a linear combination of the
$\phi_{i_1,\ldots,i_d}$. A straightforward calculation shows that

$$(2.2) \qquad [\phi_{i_1,\ldots,i_d}, X_k] = \sum_{t_1,\ldots,t_d=1}^{n} B_{k;t_1,\ldots,t_d}^{i_1,\ldots,i_d} \cdot \phi_{t_1,\ldots,t_d}$$

where the subscripts on the basis elements on the right have not been brought into order $t_1 \leq t_2 \leq \ldots \leq t_d$, and where the co-efficients have the form

$$(2.3) \quad B_{k;t_1,\ldots,t_d}^{i_1,\ldots,i_d} = c_{i_1 k t_1} \delta_{i_2 t_2} \delta_{i_3 t_3} \cdots \delta_{i_d t_d} + \ldots$$
$$+ c_{i_r k t_1} \delta_{i_1 t_2} \cdots \delta_{i_{r-1} t_r} \delta_{i_{r+1} t_{r+1}} \cdots \delta_{i_d t_d} + \ldots$$
$$+ c_{i_d k t_1} \delta_{i_1 t_2} \cdots \delta_{i_{d-1} t_d}.$$

Any Casimir invariant of degree d can be expressed as

$$(2.4) \quad C = \sum_{\substack{i_1,\ldots,i_d=1 \\ i_1 \leq \ldots \leq i_d}}^{n} P_{i_1,\ldots,i_d} \cdot \phi_{i_1,\ldots,i_d}.$$

Imposing the conditions (2.1) that C be a Casimir invariant and using (2.2) leads to

$$\sum_{\substack{i_1,\ldots,i_d=1 \\ i_1 \leq \ldots \leq d}}^{n} \sum_{t_1,\ldots,t_d=1}^{n} B_{k;t_1,\ldots,t_d}^{i_1,\ldots,i_d} \cdot P_{i_1,\ldots,i_d} \cdot \phi_{t_1,\ldots,t_d} = 0,$$

which may be written as

$$(2.5) \quad \sum_{t_1,\ldots,t_d=1}^{n} A_{k;t_1,\ldots,t_d} \cdot \phi_{t_1,\ldots,t_d} = 0,$$

where

$$A_{k;t_1,\ldots,t_d} = \sum_{\substack{i_1,\ldots,i_d=1 \\ i_1 \leq \ldots \leq i_d}}^{n} B_{k;t_1,\ldots,t_d}^{i_1,\ldots,i_d} \cdot P_{i_1,\ldots,i_d}.$$

If the subscripts on $\Phi_{t_1,\ldots,t_d}$ in (2.5) are brought into order $t_1 \leq t_2 \leq \ldots \leq t_d$ and the linear independence of the basis elements is used, the system of homogeneous linear equations that the coefficients $P_{i_1,\ldots,i_d}$ of a Casimir invariant of degree d must satisfy may be expressed simply as

$$(2.6) \qquad \sum_{\sigma \in S_d} A_{k;t_{\sigma(1)},\, t_{\sigma(2)},\ldots,\, t_{\sigma(d)}} = 0,$$

for all $k=1,\ldots,n$ and all $t_1 \leq t_2 \leq \ldots \leq t_d$.

3.   The Systems of Equations for Degrees One, Two, and Three

A. $d=1$

$$(3.1a) \qquad B_{k;t}^{\,i} = c_{ikt}$$

$$(3.1b) \qquad A_{k;t} = \sum_{i=1}^{n} c_{ikt} P_i$$

$$(3.1c) \qquad A_{k;t} = 0$$

$$(3.1d) \qquad \sum_{i=1}^{n} c_{ikt} P_i = 0, \quad k,t=1,\ldots,n$$

B. $d=2$

$$(3.2a) \qquad B_{k;st}^{\,ij} = c_{iks}\delta_{jt} + c_{jks}\delta_{it}$$

$$(3.2b) \qquad A_{k;st} = \sum_{\substack{i,j=1 \\ i \leq j}}^{n} (c_{iks}\delta_{jt} + c_{jks}\delta_{it})P_{ij}$$

$$(3.2c) \qquad A_{k;st} + A_{k;ts} = 0$$

$$(3.2d) \qquad \sum_{\substack{i,j=1 \\ i \leq j}}^{n} (c_{iks}\delta_{jt} + c_{jks}\delta_{it} + c_{ikt}\delta_{js} + c_{jkt}\delta_{is})P_{ij} = 0$$

for all $k=1,\ldots,n$ and $1 \leq s < t \leq n$.

C.   d=3

(3.3a) $B_{k;stu}^{ijr} = c_{iks}\delta_{jt}\delta_{ru} + c_{jks}\delta_{it}\delta_{ru} + c_{rks}\delta_{it}\delta_{ju}$

(3.3b) $A_{k;stu} = \sum_{\substack{i,j,r=1 \\ i\leq j\leq r}}^{n} (c_{iks}\delta_{jt}\delta_{ru} + c_{jks}\delta_{it}\delta_{ru} + c_{rks}\delta_{it}\delta_{ju})P_{ijr}$

(3.3c) $A_{k;stu} + A_{k;sut} + A_{k;ust} + A_{k;uts} + A_{k;tus} + A_{k;tsu} = 0$

(3.3d) $\sum_{\substack{i,j,r=1 \\ i\leq j\leq r}}^{n} (c_{iks}\delta_{jt}\delta_{ru} + c_{jks}\delta_{it}\delta_{ru} + c_{rks}\delta_{it}\delta_{ju}$

$+ c_{iks}\delta_{ju}\delta_{rt} + c_{jks}\delta_{iu}\delta_{rt} + c_{rks}\delta_{iu}\delta_{jt}$

$+ c_{iku}\delta_{js}\delta_{rt} + c_{jku}\delta_{is}\delta_{rt} + c_{rku}\delta_{is}\delta_{jt}$

$+ c_{iku}\delta_{jt}\delta_{rs} + c_{jku}\delta_{it}\delta_{rs} + c_{rku}\delta_{it}\delta_{js}$

$+ c_{ikt}\delta_{ju}\delta_{rs} + c_{jkt}\delta_{iu}\delta_{rs} + c_{rkt}\delta_{iu}\delta_{js}$

$+ c_{ikt}\delta_{js}\delta_{ru} + c_{jkt}\delta_{is}\delta_{ru} + c_{rkt}\delta_{is}\delta_{ju})P_{ijr}=0$

for all k=1,...,n and all $1\leq s\leq t\leq u\leq n$.

4.   Algorithms and Their Implementation

In (2.4) we want the exact values of the P's, which are to
be obtained as solutions of the system of equations (2.6).
Roundoff error can be eliminated by using integer arithmetic
throughout. Most Lie algebras of interest in applications can
be defined by rational (and thus, by multiplying each of them
by their least common denominator, integer) structure constants,
so that integer arithmetic is possible.

Algorithm A(d): For fixed degree d, generate the matrix of
coefficients of the P's in (2.6) in the following way. As each
equation is generated, its linear dependence on those previously
obtained is removed. If there is still something left, it is

entered as a row in the matrix so that its first non-zero entry
is on the main diagonal, thus obtaining an upper triangular matrix,
whose rank is the number of non-zero diagonal elements. Find the
l.c.m., say Q, of the non-zero diagonal elements and multiply the
non-zero rows by constants (namely Q/leading coefficient) so that
all non-zero diagonals equal Q. Go down the main diagonal changing
each Q to 0 and each 0 to -Q. Now the non-zero columns are linearly independent and they give the coefficients of the ϕ's as
desired in (2.4). Thus we have a basis for $ZU^{(d)}(L)$, which can
just be read off from these columns.

The program thus goes as follows:

A. Input structure constants.

B. Do Algorithm A(1), obtaining the center of L.

C. Read off center, print it out (if desired), and store for
    later use.

D. Do Algorithm A(2).

E. Calculate products of central elements and symmetrize and
    subtract out (after printing out, if desired) of the matrix
    obtained in D.

F. Read off remaining Casimirs (new ones not obtainable from
    those of lower degree) from the columns, print them out,
    and store them for later use (provided higher degree Casimir
    invariants are to be calculated).

G. Increase degree by 1 and repeat sequence analogous to Steps
    D, E, and F with the new degree. Continue until the maxi-
    mum degree desired is obtained.

5.   Concluding Remarks

The authors earlier calculated the second degree Casimir
invariants for a large number of Lie algebras of dimension
n<14 of interest in physics. This was a FORTRAN program and
was executed on the IBM 370 at Texas Tech University. This
work revealed the fact that, even though there are a large
number of equations to be considered in (3.2d), the upper

triangular matrix resulting from this system of equations is very sparse. A look at (3.3d) and, more generally, (2.5) points out how fast the size of the matrix grows with increasing values of the degree d. Thus it becomes essential to make use of the sparseness of the matrix in order to cut down on the storage required for the execution of the program. The authors are presently developing such algorithms as well as the programs to implement them.

References

[1]   E. Angelopoulos, *Sur les opérateurs de Casimir de toute algèbra de Lie*, C. R. Acad. Sc. Paris 264(1967), 585-586. MR 38 #273.

[2]   J. G. F. Belinfante and B. Kolman, *A Survey of Lie Groups and Lie Algebras with Application and Computational Methods*, SIAM, Philadelphia, 1972. MR 50 #7269.

[3]   E. G. Beltrametti and A. Blasi, *On the number of Casimir operators associated with any Lie group*, Phys. Letters 20 (1966), 62-64. MR 33 #222.

[4]   N. Jacobson, *Lie Algebras*, Interscience, New York, 1962. MR 26 #1345.

[5]   J. Patera, R. T. Sharp, P. Winternitz, and H. Zassenhaus, *Invariants of real low order Lie algebras*, J. Math. Phys. 17 (1976), 986-994.

[6]   J. Patera, R. T. Sharp, R. Winternitz, and H. Zassenhaus, *Subgroups of the similitude group of three-dimensional Minkowski space*, Canad. J. of Phys. (to appear).

[7]   J. Patera, R. T. Sharp, P. Winternitz, and H. Zassenhaus, *Subgroups of the Poincaré group and their invariants*, J. Math. Phys. 17 (1976), 977-985.

[8]   M. Pauri and G. M. Prosperi, *On the construction of the invariant operators for any finite-parameter Lie group*, Nuovo Cimento 43A (1966), 533-537.

[9]   G. Racah, *Sulla caratterizzazione delle rappresentazioni irriducibili dei gruppi semisemplici di Lie,* Atti Accad. Naz. Lincei. Rend. Cl. Sci. Fis. Mat. Nat. 8 (1950), 108-112. MR 12, page 9.

Department of Mathematics
Texas Tech University
Lubbock, Texas 79409

Division of Science and Mathematics
Edward Waters College
Jacksonville, Florida   32209

COMPUTING THE STRUCTURE OF A LIE ALGEBRA

Robert E. Beck
Bernard Kolman
Ian N. Stewart

1. Introduction

The aim of this paper is to obtain algorithms for locating
in a finite-dimensional Lie algebra L items of structural interest.
This structural analysis will include calculating the various
radicals of L, the various series of L, a Cartan subalgebra of
L, and the derivation algebra of L as well as determining whether
L is nilpotent, soluble, or semisimple.

It is surprising how much of the traditional mathematical
theory of Lie algebras is purely existential. For example,
whilst it is obvious that a maximal soluble ideal exists in L
and is unique, it is not immediately clear how it may be found
algorithmically. Again, the construction of a Cartan subalgebra
as the null component of a regular element fails to specify how
to locate a regular element. Even if algorithms for these
structural items can be found, another difficulty must be faced.
Slight changes in the structure constants of L can lead to dras-
tic alterations of its isomorphism type. For instance, let L
be a 2-dimensional Lie algebra with basis $\{x,y\}$ and suppose
$[x,y] = \lambda x$ for a scalar λ. If $\lambda = 0$ then L is abelian; if $\lambda \neq 0$
then L is isomorphic to the non-abelian soluble algebra with
basis $\{a,b\}$, and multiplication $[a,b] = a$ by the map $x \longmapsto a$,
$y \longmapsto \lambda b$. If λ is pertubed away from $\lambda = 0$ by any amount what-

soever, the isomorphism type changes. The study of this type
of question is difficult and involves techniques of deformation
theory.

Our point of view will be the following: We shall work
over a ground field F.   At times we will impose restrictions
on F, such as requiring that F be algebraically closed or of
characteristic zero.   We assume that all computations within F
may be performed exactly.

A Lie algebra L over F will be specified by giving a basis
$\{x_1, x_2, \ldots, x_n\}$ and a set of *structure constants* c^k_{ij} which
define the multiplication according to the equations

(1) $$[x_i, x_j] = \sum_{k=1}^{n} c^k_{ij} \, x_k.$$

A subspace, subalgebra, or ideal of L with basis $\{e_1, e_2, \ldots, e_r\}$
will be specified by the $r \times n$ matrix $B = [b_{ij}]$ where

$$e_i = \sum_{j=1}^{n} b_{ij} x_j \quad .$$

An L-module will be specified by giving a basis $\{p_1, p_2, \ldots, p_r\}$
and the action of L on these basis elements:

$$x_i p_j = \sum_{k=1}^{r} m^k_{ij} p_k.$$

A linear map between vector spaces will be specified by its
matrix with respect to previously specified bases.

In matters of terminology we shall follow Humphreys [4],
which we shall also use as a reference for standard theorems.

The algorithms we shall give allow us to construct most
of the fundamental objects occurring in the structure theory,
such as the nil or soluble radical, a Cartan subalgebra, a
Borel subalgebra.   Once these are known it is easy to decide
whether an algebra is semisimple, and if so to list its simple

direct summands and multiplicities and to identify their isomor-
phism types in the standard list $(A_\ell, B_\ell, C_\ell, D_\ell, G_2, F_4, E_6, E_7, E_8)$
by way of Dynkin diagrams. The methods required are often
different from the standard theorems. On occasion we use an
existential result to show the correctness of an algorithm.

In the final section we give an example to show how the
computations work out in practice.

2.   Linear Algebra

There are many standard algorithms from linear algebra
that are used in computational Lie algebra work. Among the
most useful of these are algorithms to:

(a)   Find the rank of a given m × n matrix.

(b)   Find a basis for the solution space of a
homogeneous system of equations.

(c)   Find the eigenvalues and associated eigen-
vectors of a given matrix.

(d)   Extend a linearly independent set of vectors
in a vector space L to a basis for L. To
describe this algorithm, let L be a vector
space over a field F, and let $\{u_1, u_2, \ldots, u_r\}$
be a linearly independent set of vectors in
L. By a *complementary basis* to $\{u_1, u_2, \ldots, u_r\}$
we mean a set of vectors $\{w_1, w_2, \ldots, w_{n-r}\}$ such
that $\{u_1, u_2, \ldots, u_r, w_1, \ldots, w_{n-r}\}$ is a basis for
L. Such a complementary basis may be found by
adjoining a basis $\{x_1, x_2, \ldots, x_n\}$ for L to
$\{u_1, u_2, \ldots, u_r\}$ and eliminating any x_i which
is linearly dependent on previous vectors
using matrix methods.

(e) Find a basis for the intersection of two
     given subspaces of a vector space. Let
     L be a vector space over a field F with
     given basis $\{x_1, x_2, \ldots, x_n\}$. Let U and
     V be subspaces of L with bases $\{u_1, u_2, \ldots, u_r\}$
     and $\{v_1, v_2, \ldots, v_s\}$ respectively, where
     each u_i and v_j is given as a linear com-
     bination of the x_k. To find a basis for
     U ∩ V we construct a complementary basis
     $\{w_1, w_2, \ldots, w_{n-r}\}$ to $\{u_1, u_2, \ldots, u_r\}$ and
     let

$$\pi: \ L \to U$$

     be the projection of L onto U:

$$\pi(u_i) = u_i \quad i = 1, 2, \ldots, r$$

$$\pi(w_j) = 0 \quad j = 1, 2, \ldots, n-r.$$

Now u ε U if and only if $\pi(u) = u$. Hence U ∩ V is the kernel
of the map

$$\pi\big|_V - id_V: \ V \to U$$

and this can be readily computed.

Repetition of this process allows us to find the intersection of any finite set of subspaces.

3. Fundamental Lie Algebra Properties

The structural analysis of a Lie algebra requires the
ability to easily compute certain key structural items of the
algebra. In this section we describe algorithms for calculating
these structural items; some of these algorithms will be given
in APL notation [3]. Since this notation only uses upper case
letters, we will freely change between the typical lower case

mathematical notation and upper case APL notation. For example, n and N will denote the same quantity, the dimension of the Lie algebra.

Let L be a Lie algebra over F with basis $\{x_1, x_2, \ldots, x_n\}$ defined by its structure constant array (c^k_{ij}).

(a) Basis for [U,V].

Let U and V be subspaces of L specified by the matrices B_1 and B_2. Assume B_1 is r × n and B_2 is s × n; hence the dimensions of U and V are r and s respectively. Let $\{u_1, u_2, \ldots, u_r\}$ and $\{v_1, v_2, \ldots, v_s\}$ be bases for U and V, respectively. The rs products can be represented in terms of the basis for L by the rs × n matrix

$M \leftarrow ((R \times S), N) \ \rho B1 + . \times \ 3 \ 1 \ 2 \ \lozenge C + . \times \lozenge B2$

The row space of M is [U,V]. By putting M into row echelon form we obtain a basis for [U,V].

(b) Matrix of adjoint map.

Let x ε L. By linearity, the problem of finding the matrix of ad x reduces to the problem of determining the matrix of ad x_k, k = 1,2,...,n. It is easily seen that this matrix is $C[;K;]$.

(c) Set of all derivations of L.

Let the n × n matrix $D = [d_{ij}]$ represent an arbitrary derivation of L. Using the properties of a derivation, we determine that D must satisfy

(2) $(D + . \times C) - C + . \times D = 1 \ 3 \ 2 \ \lozenge (1 \ 3 \ 2 \ \lozenge C) + . \times D$

Since (2) is linear in the entries of D,
we see that any derivation is a solution
of the homogeneous system of equations defined by (2). Specifically, there are n^3
equations in the n^2 unknowns d_{ij}.

(d) Check whether a subspace is a subalgebra or ideal.
Let U be an r-dimensional subspace of L
specified by the r × n matrix B. Using
the definition of subalgebra, we have that
U is a subalgebra of L if and only if for
all i,j 1≤ i < j ≤ r

$$r = \text{rank } (\lozenge B), M[I;;J]$$

where

$$M \leftarrow B + . \times (2\ 1\ 3\ \lozenge C) + . \times \lozenge B$$

Likewise, W is an ideal if and only if for
all i, $1 \le i \le n$ and all j, $1 \le j \le r$,

$$r = \text{rank}(\lozenge B), (C + . \times \lozenge B)\ [;I;J]$$

(e) Structure constants for L/I.
Let I be an ideal of L of dimension r
specified by the r × n matrix B. The
basis $\{e_1, e_2, \ldots, e_r\}$ for I can be
extended to a basis for L by adjoining
n-r elements from $\{x_1, x_2, \ldots, x_n\}$. The
structure constants for L/I will be given
with respect to the natural projections of
these n-r elements, $x_{i_1}, x_{i_2}, \ldots, x_{i_{n-r}}$. We
let CP be the structure constant array of
L with respect to $\{e_1, e_2, \ldots, e_r, x_{i_1}, x_{i_2}, \ldots, x_{i_{n-r}}\}$. The quotient structure constant
array CQ is given by

$$CQ \leftarrow (3\rho R - N) \uparrow CP$$

(f)   Centre of L

The *centre* of L, Z(L), is the ideal defined by

$$Z(L) = \{x \; \varepsilon \; L \; | [x,y] = 0 \text{ for all } y \; \varepsilon \; L\}.$$

Note that the property which defines the centre is linear in y. Thus, it suffices to only work with a basis for L. We find that x ε Z(L) if and only if x is a solution to the homogeneous system of equations whose matrix is

$$M \leftarrow ((N \times N), N) \rho C$$

(g)   Centralizer of a subspace

Let U be an r-dimensional subspace of L speci-fied by the r × n matrix B. The *centralizer* $C_L(U)$ is the subalgebra defined by

$$C_L(U) = \{x \; \varepsilon \; L | [U,x] = 0\}.$$

By linearity it again suffices to only consider a basis for U. We find that x ε $C_L(U)$ if any only if x is a solution to the homogeneous system of equations whose matrix is

$$M \leftarrow ((R \times N), N) \rho \; 1 \; 3 \; 2 \; \lozenge C + . \times \lozenge B$$

(h)   Idealizer (normalizer) of a subspace
Let U be as in (g). The *idealizer (normalizer)* $I_L(U)$ is the subalgebra defined by

$$I_L(U) = \{x \; \varepsilon \; L | [x,U] \subset U\}.$$

Working only with a basis for U we find that x ε $I_L(U)$ if and only if there is an r × r matrix G such that

(3) $$0 = ((2 \; 3 \; 1 \; \lozenge C + . \times \lozenge B) + . \times X - \lozenge(\lozenge B) + . \times G$$

where X denotes the coordinate vector of x with respect to the fixed basis for L. This APL expression is linear in the unknowns X and G. We write G as a vector and adjoin it to X to form the vector of unknowns

$$Y = (x_1, x_2, \ldots, x_n, g_{11}, g_{12}, \ldots, g_{1r}, g_{21}, \ldots, g_{rr})$$

Equation (3) defines a homogeneous system of equations $MY = 0$ where M is the $(nr) \times (r^2 + n)$ matrix

$$\begin{bmatrix} B^T & 0 & . & . & . & 0 & P_1 \\ 0 & B^T & & & & 0 & P_2 \\ . & & & & & . & . \\ . & & & & & . & . \\ . & & & & & 0 & . \\ 0 & & . & . & . & 0 & B^T & P_r \end{bmatrix}$$

and P_i is the $n \times n$ matrix given in APL notation by

$$PI \leftarrow (-C + . \times \Diamond B)[\,;\,;I\,]$$

Performing elementary row operations on each set of n rows of M and rearranging the rows we transform M to

$$\begin{bmatrix} I_{r^2} & Q_1 \\ \hline 0 & Q_2 \end{bmatrix} \quad ,$$

where I_r2 is the $r^2 \times r^2$ identity matrix, Q_1 is $r^2 \times n$ and Q_2 is $r(n-r) \times n$. Then $Y = \begin{bmatrix} Y_1 \\ X \end{bmatrix}$ belongs to the solution space of M if and only if X belongs to the solution space of Q_2. To find a basis for the idealizer of U we must find a basis for the solution space of Q_2. The following APL statements generate the matrix Q_2:

$$M \leftarrow (\lozenge B),(N,R \times N)\rho \ 1 \ 3 \ 2 \ \lozenge - C + . \times \lozenge B$$
$$M \leftarrow ESCH \ M$$
$$Q2 \leftarrow ((R \times (N- \ )),N)\rho (R,R) \neq M$$

The function ESCH puts a given matrix in reduced row echelon form.

4.   Series

Certain sequences of ideals play important roles in determining the structure of L.

(a)   The derived and lower central series.

Let $L^{(0)} = L$, $L^{(1)} = [L,L],\ldots,L^{(k+1)} = [L^{(k)}, L^{(k)}]$. The sequence of ideals $L^{(0)} \supset L^{(1)} \supset \ldots \supset L^{(k)} \supset \ldots$ is called the *derived series of L*. If for some k, $L^{(k)} = 0$ then L is *soluble*. Although the construction of this series could be used as a test for the solubility of L, a better criterion will be developed below. Let $L^0 = L$, $L^1 = [L,L],\ldots,$ $L^{k+1} = [L^k,L]$. The sequence of ideals $L^0 \supset L^1 \supset \ldots \supset L^k \supset \ldots$ is called the *lower central series* of L. If for some k, $L^k = 0$ then L is *nilpotent*. We will give other tests for the nilpotency of L below. Each of these series is constructed using the algorithm (Section 3(a)) for determining a basis for [U,V], where U and V are subspaces of L.

(b)   Upper central series.

Let $Z_0 = 0$, and let Z_i be the ideal of L such that

$$Z_i/Z_{i-1} = Z(L/Z_{i-1})$$

Note that $x + Z_{i-1} \varepsilon Z(L/Z_{i-1})$ if and only if $[x,y] \varepsilon Z_{i-1}$ for all $y \varepsilon L$. The ideal Z is the center of L. The sequence of ideals

$$Z_o \subset Z_1 \subset \ldots \subset Z_k \subset \ldots$$

is called the *upper central series* of L. There is a smallest integer n such that $Z_n = Z_{n+1} = \ldots$. The ideal Z_n is called the *hypercentre* of L and is denoted by Z_ω. It can be shown that L is nilpotent if and only if $Z_\omega = L$, which provides another test for nilpotency.

We shall describe the algorithm for the upper central series by giving the steps necessary to compute Z_i given Z_{i-1}. Suppose Z_{i-1} is specified by the $r \times n$ matrix B, and the multiplication table for L/Z_{i-1} is given by the $(n-r) \times (n-r) \times (n-r)$ array C'.

Step 1.   Compute the centre of L/Z_{i-1}. Its basis, in terms of the basis for L/Z_{i-1}, is given by the $k \times (n-r)$ matrix B_1.

Step 2.   Expand B_1 to a $k \times n$ matrix $\hat{B}_1$ by inserting zeroes in the colums corresponding to basis elements of L which are in Z_{i-1}.

Step 3.   Adjoin $\hat{B}_1$ to the matrix B to

obtain an $(r+k) \times n$ matrix $\hat{B}$ which specifies z_i.

Step 4. Form the array $\hat{C}'$, which specifies the multiplication table of L/Z_i.

The algorithm stops when B_1 in Step 1 is zero.

5.   The Soluble Radical

The usual definition of the soluble radical $\sigma(L)$ as the unique maximal soluble ideal of L is non-constructive. If char $(F) = 0$, then $\sigma(L) = [L,L]^{\perp}$ where the orthogonal complement is taken with respect to the Killing form. The matrix K of the Killing form can be easily computed as

$$K \leftarrow +/\ 3\ 1\ 2\ 3\ \lozenge C + . \times C$$

A basis for $[L,L]$ can be computed with the algorithm for the Lie product of two subspaces of L (Section 3a). To determine $[L,L]^{\perp}$ we first find a basis $\{e_1, e_2, \ldots, e_r\}$ for $[L,L]$ with respect to which $K\mid_{[L,L]}$ is diagonal. Extend this basis to a basis $\{e_1, e_2, \ldots, e_r, f_1, f_2, \ldots, f_{n-r}\}$ for L. We now transform this last basis for L to a basis S with respect to which K is diagonal. Necessarily $e_1, e_2, \ldots, e_r$ are the first r elements of S. Say the last n-r elements of S are $\hat{f}_1, \hat{f}_2, \ldots, \hat{f}_{n-r}$, and the diagonal elements of the corresponding matrix K for the Killing form are $\hat{k}_1, \hat{k}_2, \ldots, \hat{k}_n$. Then a basis for $[L,L]^{\perp}$ consists of $\{\hat{f}_1, \hat{f}_2, \ldots, \hat{f}_{n-r}\}$ and any e_i for which $\hat{k}_i = 0$.

6.   The Nil Radical

Again the traditional definition of the nil radical $\nu(L)$ as the unique maximal nilpotent ideal is non-constructive. We let $S = \sigma(L)$, $N = \nu(L)$. Then N is also equal to $\nu(S)$ [2, p.67]. It is known [4, p. 16] that S has a series

$$0 = S_0 \subset S_1 \subset \dots \subset S_m = S$$

where each S_i is an ideal of S and dim S_i = i, $1 \le i \le m$.

The following lemma gives a computational description of the nil radical.

Lemma 1.   The nil radical N is given by

(5) $$N = \bigcap_{i=1}^{m} C_L(S_i/S_{i-1}),$$

where $C_L (S_i/S_{i-1}) = \{x \in L \mid [x,S_i] \subset S_{i-1}\}$.

Proof. Let M denote the right-hand side of (5). Then M is an ideal of L having a *central* series $\{M \cap S_i\}_{i=1}^{m}$, which implies that M is nilpotent and $M \subset N$.   On the other hand, if $N \cap S_1 = 0$ then $[N,S_1] = 0$, so $N \subset C_L(S_1)$. If $N \cap S_1 \ne 0$ then $N \cap S_1$ intersects the centre of N nontrivially [7, p. 460], and since dim $S_1 = 1$ we have $S_1 \subset Z_1$ (N), and again $N \subset C_L(S_1)$. An easy induction shows that $N \subset M$, hence N = M.

The algorithm to construct the nil radical for a Lie algebra L over a field of characteristic zero consists of the following steps.

Step 1.   Find the soluble radical $S = \sigma(L)$ as in Section 5.

Step 2.   Construct the sequence of ideals given in (4).

Step 3.   Construct the set of centralizers $C_L(S_i/S_{i-1})$, i = 1,2,...,m using an appropriate modification of the algorithm in Section 3g.

Step 4.   Find the intersection of the centralizer using the algorithm in Section 2e.

It remains to give an algorithm for Step 2. Fortunately the method of proof of Humphreys [4, p. 16] is already algorithmic. It yields the following method for finding a basis $\{e_1,e_2,\dots,e_m\}$ for S such that $\{e_1,e_2,\dots,e_i\}$ is a basis for S_i.

Step a.   Compute $Z(S)$. If $Z(S) = 0$, go to Step c.
Otherwise, let $\{e_1, e_2, \ldots, e_r\}$ denote the basis for
$Z(S)$ which has just been computed. Thus, we have
calculated $S_1, S_2, \ldots, S_r$ in (4).

Step b.   Let S now denote $S/Z(S)$.

Step c.   Find a basis for S^2 (Section 3a).

Step d.   Extend this basis to a basis for S, say
$\{y_1, y_2, \ldots, y_{p-1}\}$. Since $S^2 \subset K$ we see that S can
be considered as a K-module under the adjoint action.

Step e.   Find a simultaneous (nonzero) eigenvector
$v \in S$ for the K-action, so that $[k,y] = \lambda(k)v$, for
all $k \in K$. We actually will not need v itself, but
do need the linear functional $\lambda: K \to F$. By linearity,
it suffices to find a simultaneous eigenvector v for
the linear maps $\text{ad } y_1$, $\text{ad } y_2, \ldots, \text{ad } y_{p-1}$, and to
specify λ by its effect on $\{y_1, y_2, \ldots, y_{p-1}\}$. We
can compute v inductively as follows.

> (i) y_1 is an eigenvector for $\text{ad } y_1$ with
> eigenvalue 0. Let $B_1 = \langle y_1 \rangle$ and $W_1 =$
> $\{w \in B_1 | \text{ad } y_1|_{B_1}(w) = \lambda(y_1)w\} = \langle y_1 \rangle$.
> (ii)   Let $B_k = \langle y_1, y_2, \ldots, y_k \rangle$. Suppose
> we have computed $W_k = \{w \in B_k | \text{ad } y_j|_{B_k}(w) =$
> $\lambda(y_j)w, \ 1 \le j \le k\}$.

We find W_{k+1} as follows. Find an eigenvalue λ_0 of
$\text{ad } y_{k+1}|_{W_k}$, and set $\lambda(y_{k+1}) = \lambda_0$. Let U_j denote the
eigenspace of $\text{ad } y_j|_{B_{k+1}}$ associated with the eigen-
value $\lambda(y_j)$, $1 \le j \le k+1$. Then

$$W_{k+1} = \bigcap_{j=1}^{k+1} U_j.$$

Thus, we have determined $\lambda(y_j)$, $1 \le j \le p-1$.

Step f.   Using the linear functional λ determined in Step e compute

$$W = \{w \in S \mid [y_j,w] = \lambda(y_j)w,\ 1 \le j \le p-1\}$$

Now W is an S-module [4, p. 16]. Thus, there is an eigenvector s ($\ne 0$) $\in$ W for ad $y_p\big|_W$. Then s spans the required ideal S_{r+1}, and we can quotient S by S_{r+1} and repeat inductively to find $S_{r+2}, \ldots, S_m$.

7.   Construction of a Non-Nilpotent Element

From now on we will denote ad x by x*. In this section we develop a procedure to find, in a non-nilpotent Lie algebra L, an element x for which x* is not nilpotent. This is a preliminary step towards the construction of a Cartan subalgebra. The difficulties arise entirely from the following fact: a Lie algebra can have a basis $\{x_1,x_2,\ldots,x_n\}$ for which all x_i* are nilpotent, and yet L is not nilpotent. Indeed, every semisimple algebra has such a basis.

To save breath, call any element x with x* nilpotent a *nilpotent element*. We know by Engel's theorem that L is nilpotent if and only if x is nilpotent for every x $\in$ L. Therefore, if L is not nilpotent, a non-nilpotent element exists. The procedure consists of the following steps:

Step 1.   If L is soluble then the set of nilpotent elements is equal to the nil radical $\nu(L)$ (Section 6). Compute $\nu(L)$ and choose x $\in$ L\$\nu(L)$. Then x is non-nilpotent.

Step 2.   If L is not soluble, compute the radical $\sigma(L) = S$. Now L/S is semisimple. If x+S $\in$ L/S is not nilpotent, then x $\in$ L is not nilpotent. Hence, without loss of generality we may assume that L is semisimple.

Step 3. Now assuming L is semisimple, we construct a sequence of subalgebras as follows.
Choose a nonzero element y of L, and let

$$Y_1 = \langle y \rangle,$$

$$Y_{i+1} = I_L(Y_i).$$

Then

$$Y_1 \subset Y_2 \subset Y_3 \subset \ldots \subset Y_c = Y_{c+1}$$

for some $c \leq$ dim L. Two cases arise.

Case 1. Every Y_i is nilpotent. Then, since
$Y_c = Y_{c+1}$, it follows that $Y_c = I_L(Y_c)$ is a
Cartan subalgebra of L. Assuming char (F) = 0,
since L is semisimple, Y_c is a maximal torus of
L [4 p. 80]. It follows that every $x \, \varepsilon \, Y_c$ is
semisimple (that is, x* is diagonalizable) and
hence, non-nilpotent since Z_1 (L) = 0. In particular, the original element Y must be non-
nilpotent.

Case 2. Some Y_i is not nilpotent. We may
take the first i with this property, in which
case Y_{i-1}, which is an ideal of Y_i, is nilpotent. We cannot have $Y_i = L$ since L, being
semisimple, has no nonzero nilpotent ideals.
Hence, dim $Y_i <$ dim L. Since Y_i is not nilpotent we may return inductively to Step 1,
replacing L there by Y_i. We will eventually
locate a non-nilpotent element in some quotient
of some subalgebra of L. Reversing the quotient
steps one can easily derive a non-nilpotent element of L.

8.   The Cartan Subalgebra

The problem of finding a Cartan subalgebra of an arbitrary
Lie algebra can be reduced to the problem of finding a Cartan
subalgebra of a semisimple algebra and of a soluble algebra.
We first describe the reduction and then give algorithms for
the two special cases.

Let L be an arbitrary Lie algebra. Let S be its soluble
radical. Then L/S is semisimple and we find a Cartan subalgebra
C/S of L/S, where C ⊃ S.

Now C is soluble, and we find a Cartan subalgebra H of C.
But now by Barnes and Gastineau-Hills [1, p. 344] (the proof
extends without trouble to the insoluble case) it follows that
H is a Cartan subalgebra of L, which completes the reduction
procedure.

Semisimple case: We first assemble several results which
will be needed in describing the algorithm. Let L be semisimple.
Then L admits a unique *Jordan-Chevalley decomposition*, namely
if $x \in L$ then there exist unique elements $x^s, x^n \in L$ such that

$$x = x^n + x^s,$$

$$0 = [x^n, x^s],$$

$$x^n* \text{ is nilpotent,}$$

$$x^s* \text{ is semisimple (diagonalizable).}$$

We can calculate the Jordan form of the linear map $x*: L \to L$.
This yields maps $n: L \to L$ and $s: L \to L$ such that $ns = sn$,
$x* = n+s$, n is nilpotent, s is semisimple. Now x^n is the unique
element of L such that

(6) $x^n* = n$

[4, p. 24]. If we write x^n in terms of the basis for L

$$x^n = \Sigma_i \lambda_i x_i,$$

then (6) is equivalent to the linear system of equations

$$\sum_i \lambda_i [x_i, x_j] = n x_j$$

Substituting from (1) allows us to solve for the λ_i since the theory already implies that a solution exists and is unique. Similarly we can compute x^s.

Assuming char $(F) = 0$, a subalgebra C is a Cartan subalgebra if and only if it is a maximal torus [4, p. 80]; further, every torus is abelian.

Lemma 2.   If L is semisimple, T a torus of L, and $C = C_L(T)$ is nilpotent, then C is a Cartan subalgebra of L.

Proof.   Let $x \in I_L(C)$ and put $X = \langle x \rangle + C$. Then X/C is a T-module, annihilated by T. Looking at the weight-space decomposition of X as T-module, we see that this implies $X \subset C$. Hence, $x \in C$, so $C = I_L(C)$ and C is a Cartan subalgebra.

The algorithm for computing a Cartan subalgebra of a semisimple Lie algebra L then proceeds as follows.

  Step 1.   Find a non-nilpotent element x in L
  (Section 7). Then x^s is a nonzero semisimple
  element and spans a torus T_0.

  Step 2.   If $C = C_L(T_0)$ is nilpotent, then C
  is a Cartan subalgebra of L by Lemma 2.

  Step 3.   If C is not nilpotent, then $T_0 \subset Z(C)$
  and C/T_0 is not nilpotent. Thus, C/T_0 contains
  a non-nilpotent element $y + T_0$. That is, $y \in C$
  and y is not nilpotent modulo T_0. Hence $y^s \notin T_0$.
  Since $y^s \in C$ [4, p. 36], $T_1 = T_0 + \langle y^s \rangle$ is a
  larger torus. We replace T_0 by T_1, form $C_L(T_1)$
  and check its nilpotency. After repeating Step 3
  at most dim L times we find a nilpotent centralizer

$C_L(T_c)$, for some c, which is a Cartan subalgebra
by Lemma 2 and the process terminates.

Soluble case: Assume L is soluble.

Step 1.   If L is nilpotent, then L is its own
Cartan subalgebra.

Step 2.   If L is not nilpotent, find a one-
dimensional ideal A of L (Section 6, Step 2).

Step 3.   If L/A is not nilpotent then, induc-
tively, we may find a Cartan subalgebra C/A of
L/A with dim C < dim L.   Then we use induction
again to find a Cartan subalgebra H of C.   Now
H is a Cartan subalgebra of L by Barnes and
Gastineau-Hills [1, p. 344].

Step 4.   If L/A is nilpotent and L is not
nilpotent we compute the hypercentre $Z = Z_\omega$
(Section 4b).   Then L/Z is not nilpotent and
has trivial centre.   If H/Z is a Cartan subalgebra of L/Z, then H is a Cartan subalgebra
of L, so we may assume inductively that Z = 0
and L has trivial centre.

Step 5.   Compute $C_L(L/A)$ and $C_L(A)$.

Step 6.   Find an element $x \in C_L(L/A)$, $x \notin C_L(A)$.
Then $P = C_L(x)$ is a Cartan subalgebra of L.

There are two points in Step 6 that need justification.
We know on theoretical grounds that there exists a Cartan subalgebra K of L.   If x is a nonzero element of Z(K), then
$x \in C_L(L/A)$, and $x \notin C_L(A)$ since $x \in Z(L)$.   This means that we
can always find the required element in Step 6.   The following
lemma shows that P as defined above is a Cartan subalgebra of L.

Lemma 3.   Let L be a soluble but not nilpotent Lie algebra with trivial hypercentre and, hence, trivial centre. Let A be a one-dimensional ideal such that L/A is not nilpotent. Let $x \in C_L(L/A) \backslash C_L(A)$. Then $P = C_L(x)$ is a Cartan subalgebra of L.

Proof.   We show that P is nilpotent and self-idealizing. The map x* sends L into A (since $x \in C_L(L/A)$) and is linear. Since $P = \ker x^*$, and dim A = 1, then dim P ≥ (dim L)-1. But if dim P = dim L we have $x \in Z(L)$, a contradiction. Hence, dim P = (dim L)-1. Now $A \not\subseteq P$ since $x \not\in C_L(A)$, so $A \cap P = 0$ and L is the split extension A ∔ P. It follows that P is nilpotent, being isomorphic to L/A. Further, P is self-idealizing, for if $P \ne I_L(P)$ then $P \subsetneq I_L(P)$. Since dim P = (dim L)-1, $I_L(P) = L$, and P is an ideal of L. Therefore, the split extension L = A ∔ P is a direct sum $L = A \oplus P$ of nilpotent ideals, hence, nilpotent, a contradiction.

9.   An Example

We illustrate these methods on a relatively simple example.

Consider the 6-dimensional Lie algebra whose multiplication is defined by

$$[x_1,x_5] = x_3, \quad [x_2,x_6] = x_1, \quad [x_3,x_4] = x_2$$

$$[x_4,x_2] = x_3, \quad [x_4,x_5] = x_6, \quad [x_5,x_3] = x_1$$

$$[x_5,x_6] = x_4, \quad [x_6,x_1] = x_2, \quad [x_6,x_4] = x_5.$$

All other products are zero. The only pairs of basis elements for which the Killing form is not zero are

$$(x_4,x_4) = (x_5,x_5) = (x_6,x_6) = -4.$$

We see that $L^{(1)} = L^1 = [L,L] = L$, so that L is neither soluble nor nilpotent. The subspace spanned by $\{x_1,x_2,x_3\}$ is an ideal so that L is not simple. We find

$$Z = 0$$

$$\sigma(L) = <x_1, x_2, x_3>$$

$$\nu(L) = <x_1, x_2, x_3>.$$

The algorithm for a Cartan subalgebra tells us to begin by finding a Cartan subalgebra for $L/\sigma(L)$, which has a basis $[x_4', x_5', x_6']$ and multiplication $[x_4', x_5'] = x_6'$, $[x_4', x_6'] = -x_5'$, $[x_5', x_6'] = x_4'$ where dashes indicate images under the natural map.

We begin by finding a non-nilpotent element. Putting $Y_1 = <x_4'>$, we find that $Y_2 = I_L(Y_1) = <x_4'> = Y_1$ also. Since Y_1 is nilpotent it follows that x_4' is a semisimple element. Now $C_{L/\sigma(L)}(Y_1) = <x_4'>$, and it follows that $<x_4'>$ is a Cartan subalgebra for $L/\sigma(L)$. Hence, in the reduction procedure we have

$$C = <x_1, x_2, x_3, x_4>$$

and $C/\sigma(L)$ is a Cartan subalgebra for $L/\sigma(L)$. Then a Cartan subalgebra of C must be a Cartan subalgebra of L. Now

$$C^2 = <x_2, x_3> = C^3$$

and

$$C^{(2)} = 0,$$

so that C is soluble but not nilpotent. We are now at Step 2 of the soluble case. Using Step 2 of the algorithm in Section 6, we find a 1-dimensional ideal $<x_1>$. Forming $C_1 = C/<x_1>$ we obtain a soluble but not nilpotent algebra. To find a Cartan subalgebra of C, we must find a Cartan subalgebra of C_1. We return to Step 2 of Section 6, and find a 1-dimensional ideal of C_1 spanned by $x_2' + i\, x_3'$. Let $C_2 = C_1/<x_2' + ix_3'> = <x_3'', x_4''>$, where $[x_3'', x_4''] = -ix_3''$. Since C_2 is again soluble and not nilpotent we again find a 1-dimensional ideal of C_2, namely $<x_3''>$.

However, this time $C_2/\langle x_3''\rangle$ is nilpotent, in fact abelian. We compute the hypercentre Z_ω of C_2, which turns out to be 0. We next compute (Step 5)

$$C_{C_2}(C_2/\langle x_3''\rangle) = C_2$$

$$C_{C_2}(x_3'') = \langle x_3''\rangle.$$

We pick an element in $C_2\setminus\langle x_3''\rangle$, namely x_4''. Thus, $\langle x_4''\rangle$ is a Cartan subalgebra for C_2. Backtracking one stage we get $K_1 = \langle x_2' + ix_3', x_4'\rangle$, a 2-dimensional soluble subalgebra of C_1. Within K_1 lies a Cartan subalgebra of C_1. Since K_1 is soluble but not nilpotent we find a 1-dimensional ideal $\langle x_2' + ix_3'\rangle = A$ of K_1. We find that

$$C_{K_1}(K_1/A) = K_1 \quad \text{and} \quad C_{K_1}(A) = A.$$

Hence, $H_1 = C_{K_1}(x_4') = \langle x_4'\rangle$ is a Cartan subalgebra of K_1. Then $\langle x_1, x_4\rangle$ contains a Cartan subalgebra of C. Since $\langle x_1, x_4\rangle$ is abelian, it is a Cartan subalgebra of C, and hence of L.

Acknowledgement

The work of Robert E. Beck and Bernard Kolman was partially supported by the United States Air Force Office of Scientific Research. Thanks are due to Joseph Heyse for his programming assistance.

References

[1] D.W. Barnes and H.M. Gastineau-Hills, *On the theory of soluble Lie algebras*, Math. Z. <u>106</u>(1968), 343-354.

[2] N. Bourbaki, *Groupes et Algèbres de Lie, Chap. I*, Hermann, Paris, 1971.

[3]   Leonard Gillman and Allen J. Rose, *APL An Interactive Approach*, 2nd edition, Wiley, New York, 1974.

[4]   James E. Humphreys, *Introduction to Lie Algebras and Representation Theory*, Springer-Verlag, New York, 1972.

[5]   Nathan Jacobson. *Lie Algebras*, Interscience, New York, 1962.

[6]   A. I. Mal'cev, *Solvable Lie algebras*, Amer. Math. Soc. Transl. Ser I 9 (1962), 229-262.

[7]   E. Schenkman, *A theory of subinvariant Lie algebras*, Amer. J. Math. 73(1951), 453-474.

Departments of Mathematics

Villanova University
Villanova, Pennsylvania   19085

Drexel University
Philadelphia, Pennsylvania   19104

University of Warwick
Coventry, England

WHAT IS THE TYPICAL NILPOTENT LIE ALGEBRA?

Eugene M. Luks

1.   Introduction

We must remark at the outset that we shall not attempt an exact definition of 'typical' much less a precise answer to the title question. Rather we shall describe some computer assisted studies of what may be thought of as 'random samples' of nilpotent Lie algebras. The motivation for these efforts was evidence that some counterexamples should, if they exist at all, be quite abundant. In fact, it is suggested that the use of the computer searches would have facilitated the answers to questions on the existence of a number of interesting classes of algebras. Actually, the answers, where they had already been found, had required laborious hand computations and considerable time.

We discuss nilpotent Lie algebras (Section 4) and, separately, the subclass of metabelian Lie algebras (Section 3). In each instance we describe a procedure which, in some sense, corresponds to the choice of a random algebra. Since there does not appear to be a natural probability measure to assign to these classes, we are guided somewhat by intuition in establishing the sampling process. We neither insist nor desire that all algebras be chosen with equal likelihood, only that, theoretically, none are specifically excluded. For this purpose, we describe certain subvarieties of the variety Lie(X) of Lie multiplications on a vector space X (see, for example, [10]). These subvarieties cut across all relevant isomorphism classes. However, unlike the full variety, their structure is suggestive of a selection procedure.

It is no surprise, in each instance, that the procedure turns out to be prejudiced in favor of the points in Lie(X) which give rise to the smallest derivation algebras, for having many independent derivations is an algebraic condition on the coordinates of the point. Indeed, this observation inspired both the nilpotent and metabelian studies since, in each case, we wanted to uncover algebras with 'few' derivations.

2.   Notations and Preliminaries

For a Lie algebra, L, we denote by Der(L) the derivation algebra of L. If L has integral structure constants we often consider the algebra, L_p, obtained by passing to the finite field Z_p. Since the rank of the system of linear equations satisfied by derivations cannot increase upon passage modulo p we have the elementary but quite useful

Proposition 2.1. $\dim(\text{Der}(L)) \leq \dim(\text{Der}(L_p))$.

Typically, we need to compute bases of derivation algebras and so we restrict our attention to algebras with integral structure constants. However, we still face computational difficulties since the integral row reduction procedures are very slow and they may lead to integer overflows on the computer. If we pass to Z_p these problems are eliminated. Fortunately, it is often the case that we can already find $\dim(\text{Der}(L_p))$ independent derivations of L, either by inspection of L itself or by pulling back elements of $\text{Der}(L_p)$. In such instances the proposition assures that these derivations are a basis of Der(L).

All vector spaces and algebras are assumed to be finite dimensional. Unless otherwise indicated $V \oplus W$, for subspaces V, W of an algebra, shall denote the direct sum as vector spaces.

3.   Metabelian Lie Algebras

3.1 Computing Derivations

In this section we describe a particularly effective algorithm for computing the derivations of metabelian Lie algebras. It makes strong use of the known structure of the derivation algebras and allows one to deal with larger algebras than would be otherwise tractable.

If L is a metabelian Lie algebra (i.e., $L^3 = 0$) and U is any linear complement in L of the derived algebra, then $L = U \oplus V$ where $V = L^2 = U^2$ and the multiplication in L is determined by the map $\mu: U \wedge U \to V$ such that $\mu(u_1 \wedge u_2) = [u_1, u_2]$. There is an injection

$$\text{Hom}(U,V) \xrightarrow{\ i\ } \text{Der}(L)$$

in which $i(f)(u) = f(u)$ for $u \in U$ and $i(f)(V) = 0$. The image of i is an abelian ideal in Der(L) over which Der(L) splits. Namely

(3.1) $\text{Der}(L) = \text{Der}_U(L) \oplus i(\text{Hom}(U,V))$

in which we have denoted by $\text{Der}_U(L)$ the subalgebra of derivations which stabilize U. Thus we only need to compute the elements of Hom(U,U) which extend to derivations of L. For this purpose, we fix bases $\{u_1, u_2, \ldots, u_m\}$, $\{v_1, v_2, \ldots, v_n\}$ of U,V, respectively. The multiplication μ is represented by n skew-symetric $m \times m$ matrices, $A^k = (A^k_{ij})$, $k = 1, 2, \ldots, n$, where

$$\mu(u_i \wedge u_j) = \sum_k A^k_{ij} v_k.$$

We let A denote the linear span of $A^1, A^2, \ldots, A^n$. Let $\tau \in \text{Hom}(U,U)$ and suppose T is the matrix of τ relative to $\{u_i\}$. Then as noted in [6, Proposition 4.7], τ extends to a derivation of L if and only if A is stabilized by the map

(3.2) $X \to TX + XT^t.$

The next step is to translate into some computable form the condition that a subspace is stabilized. One procedure is to choose a reasonable complement to A within the space, Sk, of all $m \times m$ skew-symmetric matrices and express the *linear* condition that $TA^i + A^iT^t$ have no component within that complement. For this purpose, we use the Killing form $<,>$ of the natural representation of $gl(m)$, i.e., $<X,Y> = tr(XY)$. We assume henceforth that the characteristic of the base field is different from 2. This assures, in particular, that the restriction of $<,>$ to Sk is non-degenerate. Let B denote the orthogonal complement of A in Sk relative to $<,>$. Then, for $A \in A$, $TA + AT^t \in A$ if and only if $<TA + AT^t, B> = 0$. However, for $A,B \in Sk$,

$$<TA + AT^t, B> = tr(TAB) + tr(AT^tB) = tr(TAB) + tr(BTA) = 2tr(TAB)=$$

$2<T,AB>$. Hence A is stabilized by the map (3.2) if and only if $<T,AB> = 0$ for $A \in A$, $B \in B$. We have

<u>Proposition 3.1</u>   Suppose L, U, A, $B=A^\perp \cap Sk$ are as above. Then $(AB)^\perp$ is a subalgebra of $gl(m)$ isomorphic to $Der_U(L)$ (where $\perp$ denotes the orthogonal complement in $gl(m)$ relative to $<,>$).

Thus, given structure constants (A_{ij}^k) of the metabelian Lie algebra L, the computation of $Der(L)$ involves two steps:

(I)   Compute a basis $B^1, B^2, \dots, B^{\binom{m}{2}-n}$ of B. This involves solving a system of n homogeneous linear equations in $\binom{m}{2}$ unknowns (the coefficients being, e.g., the above-diagonal entries of the A^i).

(II) Compute a basis of $(\text{span } \{A^i B^j\})^\perp$. This involves solving $n(\binom{m}{2}-n)$ equations in m^2 unknowns.

Note that a direct computation of $Der(L)$ would require solving a system of $\binom{m+n}{2}(m+n)$ equations in $(m+n)^2$ variables. Even if one restricts one's attention to $Der_U(L)$, the derivations stabilizing U and V, direct computation involves solving a system with m^2+n^2 unknowns.

3.2   The 'Typical' Metabelian Lie Algebras

Before indicating the method for drawing an element at random from the urn of metabelian Lie algebras, it is useful to discuss what we were hoping to find therein.

Suppose $L = U \oplus V$, A, B are as in Section 3.1. Since $\text{tr}(IAB) = \text{tr}(AB) = 0$ for $A \in A$, $B \in B$, $(AB)^{\perp}$ always contains the identity matrix, I. From this, or directly, one sees that the identity map of U extends to a derivation of L. In [6] and [7] the author and G. Leger considered the class of metabelian Lie algebras in which this derivation actually spans $\text{Der}_U(L)$. We shall say that $L = U \oplus V$ is of type $I(m,n)$, where $m = \dim(U)$, $n = \dim(V)$, (or, simply, of type I if it is not necessary to explicate m,n) if $\dim(\text{Der}_U(L)) = 1$. One of the reasons that such algebras appeared to be of interest is that their holomorphs seem to forget the multiplication of the algebra. Recall that the holomorph, Hol(L), of a Lie algebra, L, is the semi-direct sum $\text{Der}(L) + L$. It is shown in [6] that Hol(L) for L of type I takes a remarkably simple form. In particular,

Proposition 2.2   (Leger-Luks [6, Section 4]). If L_1 and L_2 are both of type $I(m,n)$ then $\text{Hol}(L_1) \cong \text{Hol}(L_2)$.

This proposition provided the key to the resolution of the question of whether nilpotent Lie algebras are determined by their holomorphs. The problem had been suggested in [11] where it was shown that the free nilpotent algebras are so determined. However, examples are exhibited in [7] of two metabelian Lie algebras over Q of type $I(6,4)$ which are non-isomorphic even under extensions of the base field. We shall return to this point in Section 3.3.

Initially, instances of metabelian Lie algebras of type I did not seem to be easy to find. In retrospect, the problem was that, because of the difficulties of hand computation, we

tended to try algebras whose structure constants were too simple
and these did not exhibit the right properties. In fact, it now
appears that truly arbitrary choices of algebras would invari-
ably uncover examples of type I. To elaborate, we consider what
may be called the varieties of metabelian Lie multiplications.

We assume vector spaces U, V are fixed and set $L = U \oplus V$.
Denote by Meta(U,V) the variety of skew multiplications μ on
L for which $\mu(L,L) \subset V$ and $\mu(V,L) = 0$. For such μ, the corresponding algebra, L_μ is a metabelian Lie algebra and $L_\mu^2 = U^2 \subset V^2$
(we do not insist the inclusion be an equality). When the
spaces do not require specification, we denote Meta(U,V) by
Meta(m,n) where m = dim(U), n = dim(V). Note that Meta(U,V)
may be identified with $\mathrm{Hom}(U \wedge U, V)$ and so Meta(m,n) is an affine
space of dimension $\binom{m}{2}n$.

By equation (3.1), if $\mu \in$ Meta(m,n) then L_μ is not of type
$I(m,n)$ if and only if the dimension of $\mathrm{Der}(L_\mu)$ exceeds $1 + \binom{m}{2}n$.
Since that is an algebraic condition on μ, we have

<u>Proposition 3.3.</u>   Suppose the base field is infinite. If the
class of algebras of type $I(m,n)$ is non-empty then $\{\mu \in$ Meta(m,n)$|$
L_μ is of type $I\}$ is Zariski-open and dense in Meta(m,n).

The proposition suggests, in effect, that if algebras of
type I exist, then they are almost all that exist. Thus some
random sampling of elements of Meta(U,V) ought to either reveal
their existence or provide strong evidence for their non-exist-
ence. The computer implementation of this admittedly imprecise
notion was irresistible.

Now, Proposition 3.3 also guarantees the existence and
density of rationally defined algebras of type I if they ever
appear in characteristic 0. Hence we may, with confidence,
restrict our attention to algebras with integral structure constants A_{ij}^k (see Section 3.1). Of course, it is not immediately

apparent how to select random integers. Furthermore, we know that if they get too large, or even if they are small but m and n are large, the row-reduction might "blow up". With this in mind, we agree in advance that we shall be passing to Z_p after generating the algebra and hope that the remarks of Section 2 can be used to determine Der(L). Indeed Proposition 2.1 and equation (3.1) yield

Proposition 3.4. Suppose L is a metabelian Lie algebra with integral structure constants. Then, if L_p is of type I, so is L.

One expects further, that if p is very large, L_p is likely to be of type I when L is. Thus, although Z_p is finite and so the probability of success each time is no longer 1, if algebras of type I exist then we still ought to find them. Hence, we *pick* our "random" A_{ij}^k from the urn: $\{0,1,...p-1\}$.

This procedure was used to search for examples employing several primes up to 46337. (That limit was imposed to guarantee that integral computations such as rs + t would not result in numbers exceeding the single precision ceiling of 2^{31}). Before reporting the result we remark that we already knew:

Proposition 2.5 [6, Theorem 4.9]. There are no algebras of type I(m,n) unless

$$m = 5 \quad \text{and} \quad 4 \leq n \leq 6$$

or

$$m \geq 6 \quad \text{and} \quad 3 \leq n \leq \binom{m}{2}-3.$$

The computer search has suggested:

Algebras of type I(m,n) do exist for (m,n) as above with the exception of (6,3) and (6,12). (Values of m up to 15 were tried).

We remark that, although we have not succeeded in *proving* that algebras of type I(6,3) and I(6,12) do not exist, the computer has given us some confidence in the conjecture.

3.3   The Abundance of Nonisomorphic Metabelian Lie Algebras With Isomorphic Holomorphs

It is worth noting that the very existence of algebras of type $I(m,n)$ could have sufficed to establish the existence of nonisomorphic metabelian Lie algebras with isomorphic holomorphs. For, by Propositions 3.2, 3.3, if metabelian Lie algebras were determined by their holomorphs then the isomorphism class of any algebra of type $I(m,n)$ is dense in Meta(m,n). However, one sees that, for most U,V, the isomorphism class of any $\mu \in$ Meta$(U,V) \equiv$ Hom$(U \wedge U, V)$, i.e., the orbit of μ under the action of

$$G = GL(U) \times GL(V)$$

cannot get that big. We assume, for the moment, that the base field is algebraically closed and of characteristic 0. Then, since the stability group of G corresponding to μ (equivalently $\{\sigma \in \text{Aut}(L_\mu) \mid \sigma(U) = U\}$) has dimension = dim$(\text{Der}_U(L))$,

<u>Proposition 3.6.</u>   The dimension of the orbit of μ is

$$(\dim U)^2 + (\dim V)^2 - \dim(\text{Der}_U(L)).$$

We remark that his proposition was suggested by results in [4, especially Section 7] where an equivalent formulation of varieties of metabelian Lie algebras is presented.

The proposition implies, in particular, that the isomorphism class of any $\mu \in$ Meta(m,n) lies in a subvariety of dimension $\leq m^2 + n^2 - 1$. Thus, by Proposition 3.3.

<u>Corollary 3.7.</u>   Suppose $m^2 + n^2 - 1 < \binom{m}{2}n$. Then, if there exist algebras of type $I(m,n)$, there exist infinitely many non-isomorphic algebras in Meta(m,n) with the same holomorph.

Now it is further noted in [4, Theorem 7.8] that, for the values of (m,n) given in our Proposition 3.5, the inequality of the corollary holds except when $(m,n) = (5,4)$ or $(5,6)$. One interpretation, then, of this discussion is that: For many values of (m,n), if *two* elements of Meta(m,n) are chosen at random then

(i) With probability 1, they are nonisomorphic.

(ii) With probability 1, their holomorphs *are* isomorphic.

In other words, holomorphs are particularly *ineffective* invariants for metabelian Lie algebras.

We remark, finally, that in the exceptional cases (5,4) and (5,6) algebras of type $I(m,n)$ do exist. Two examples of algebras of type $I(5,4)$ were given in [6, Section 4] which we originally had hoped were nonisomorphic (they *are* nonisomorphic over Q). It is interesting that, although we have been unable to exhibit the actual isomorphism, the fact that both have dense orbits implies that they *are* necessarily isomorphic over C.

3.4 An Answer to a Question on the Existence of Dense Orbits

We conclude the discussion of metabelian Lie algebras with a computer-inspired answer to a question of Gauger (see [4, remark (1) on page 326]).

In the setting of the present paper, the question is whether the existence of an element μ whose orbit under $GL(U) \times GL(V)$ is dense in Meta(U,V) implies that the total number of orbits (i.e., isomorphism classes) is finite. A search for a counter-example was inspired by the result [4, Theorem 7.10] that there are an infinite number of isomorphism classes of algebras in Meta(m,2) for $m \geq 8$. Can one, nevertheless, be dense?

Noting Proposition 3.6, we search for an element of Meta(U,V) = Meta(m,2) with $\dim(\text{Der}_U(L)) = m^2 + 2^2 - \binom{m}{2}2 = m + 4$. Again, a "random" choice ought to find such an element, if it exists. For m = 8, the search was futile, but for m = 9, the computer found one every time. With this assurance, we went on to look for one with 'simple' structure constants. For example, the following $\mu \in \mathrm{Hom}(U \wedge U, V)$ works:

$$\mu(u_1 \wedge u_2) = \mu(u_3 \wedge u_4) = \mu(u_5 \wedge u_6) = \mu(u_7 \wedge u_8) = v_1,$$

$$\mu(u_1 \wedge u_9) = \mu(u_2 \wedge u_3) = \mu(u_4 \wedge u_7) = \mu(u_6 \wedge u_8) = v_2$$

with $\mu(u_i \wedge u_j) = 0$ for $i < j$, otherwise.

4.   Nilpotent Lie Algebras

4.1   Generating 'Random' Algebras

We indicate, in this section, the procedure for generation of nilpotent Lie algebras. Recall first that a nilpotent Lie algebra, L, of dimension n has a central composition series

$$L = V_0 \supset V_1 \supset \ldots \supset V_n = 0$$

of ideals V_i, i.e., $\dim V_i = n-i$ and $[V_i, L] \subset V_{i+1}$. Then L/V_{i+1} is a central extension of L/V_i by a one-dimensional ideal and so it is determined by L/V_i and an element of $Z^2(L/V_i)$, that is, a 2-cocyle with trivial coefficients.

With the above in mind, we suppose now that

$$V: \ 0 = V_n \subset V_{n-1} \subset \ldots \subset V_0$$

is a fixed tower of vector spaces with $\dim V_i = n - i$. Set $L = V_0$ and denote by $\mathrm{Nilp}(V)$ the variety of Lie multiplications μ on L for which $\mu(V_i, L) \subset V_{i+1}$. Our selection of an arbitrary point in $\mathrm{Nilp}(V)$ is accomplished by choosing, for each i, a multiplication on L/V_{i+1} which extends that previously determined on L/V_i. These multiplications are determined by successive choices of 2-cocycles. Thus, at each stage, we choose a 'random' solution to the system of linear homogeneous equations determined by the cocycle condition. This is done by row-reducing the system and assigning random values to the free variables. (When working over the integers, it may be necessary to mutiply the solution by an appropriate least common denominator to clear the fractions in the bound variables).

Note that the selection, by chance, of a coboundary would result in $L/V_{i+1} = (x) \oplus L/V_i$ (algebra direct sum) and we do not specifically exclude the possiblity. However, it is known that, for N nilpotent of dimension ≥ 2, $H^2(N) \neq 0$ (see e.g., [1]) and so, if the base field is infinite, the selection of coboundaries ought to occur with probability 0. Indeed, except in the case of small finite fields, coboundaries were never chosen, with the result that the nilpotent L always had two generators. To study, realistically, the nilpotent algebras with

$$\dim(L/L^2) = r > 2$$

one may specify the first r cocycle selections to be 0.

We further comment that there are undoubtedly many other important classes of algebras against which our procedure is biased. For example, in dimensions ≥ 7 we never seemed to find algebras of maximal nilpotency index, dim L. However, even though such multiplications form an open set in $\text{Nilp}(V)$, the result is not very surprising for there is evidence that these lie in components of relatively small dimensions (see [12, Introduction]).

Remark. We observe that the generation procedure can be generalized to select super-solvable algebras, i.e., those having composition series with one-dimensional quotients. The only difference is that one does not assume the successive extensions are central. Thus, given $K = L/V_i$, we first choose a random one-dimensional representation, ρ, of K, that is, an element of $(K/K^2)*$, and then a random element of $Z^2(K,\rho)$.

4.2   Characteristically Nilpotent Lie Algebras

As in the metabelian case, we had some idea of the properties to expect in the typical nilpotent algebra although the situation is not quite so clear cut.

Now in [5], Jacobson showed that over a field of characteristic 0, if a Lie algebra has a nonsingular derivation then it
is necessarily nilpotent. He then asked whether, in fact, every
nilpotent Lie algebra actually has a nonsingular derivation.
Quite to the contrary, it was found

Proposition 4.1   (Dixmier-Lister [2]). There exists an 8-
dimensional nilpotent Lie algebra with only nilpotent derivations.

Dixmier and Lister called algebras with this property
characteristically nilpotent and they remarked that the study
of these "might prove more tractable" than that of the larger
class of general nilpotent Lie algebras. The object of our
computer search was to see how large the subclass was. In
fact, because the typical algebra should have the fewest derivations we expected to get characteristically nilpotent algebras very often, at least in dimensions ≥ 7. (Morosov's classification [9] demonstrates that no characteristically nilpotent
Lie algebras exist in characteristic 0 through dimension 6. On
the other hand there is one of dimension 7 [3]). To be sure
having *few* derivations is not the same as having only nilpotent
derivations but there is evidence of some correlation between
the two properties. For example, any linear transformation of
L mapping L to $L^2 \cap$ Center(L) and L^2 to 0 is a nilpotent derivation.
Indeed, Dixmier and Lister showed that the outer derivation algebra of their example is already spanned by the images of such
derivations. This inspires their remark that the algebra "has
as few other derivations as possible." In a sense, the metabelian
situation lends further evidence. Of course, metabelian Lie
algebras cannot quite be characteristically nilpotent since, by
virtue of the relation $L^3 = 0$, Der(L) inherits a derivation which
induces the identity map on L/L^2. However, by Proposition 3.3,
the 'typical' metabelian Lie algebra comes as close to character-

istic nilpotence as it can get.

Thus motivated, we generated nilpotent Lie algebras according to the procedure in Section 4.1 and computed their derivation algebras. The size of the algebras was limited (we went to dimension 15) by the need to solve equations in $(\dim L)^2$ unknowns for the derivations. We found

The algebras of dimension $\geq$ 8 were always characteristically nilpotent.

(We comment on the dimension 7 case in Section 4.4).

There was no difficulty in verifying the nilpotence of all derivations since the computed basis of the derivation algebra was always strictly triangular with respect to the given basis (i.e., $\{x_1, x_2, \ldots, x_n\}$ where $\{x_{i+1}, \ldots, x_n\}$ is a basis of V_i) of L. For computational convenience, once again, most of the algebras were generated over large finite fields. However, it was reasonable, having confidence now in their abundance, to seek characteristic 0 examples on the computer. Although we had to take care now to pick only 'small' solutions to the cocycle equations, we often succeeded. We present here one of the computer generated examples: L has basis $\{x_1, x_2, \ldots, x_8\}$ and multiplication determined by

$$[x_1, x_2] = x_3 - x_4 + x_5 + x_7$$

$$[x_1, x_3] = x_4 + x_5 - x_6 - x_7 - 2x_8$$

$$[x_1, x_4] = -x_5 - x_6 + x_7$$

$$[x_1, x_5] = x_6 + x_8$$

$$[x_1, x_6] = x_8$$

$$[x_1, x_7] = -2x_8$$

$$[x_2, x_3] = -x_5 - 2x_6 + x_8$$

$$[x_2, x_4] = -3x_6 - 2x_7 + x_8$$

$$[x_2, x_5] = x_6 + x_7$$

$$[x_2, x_6] = -2x_8$$

$$[x_2, x_7] = 2x_8$$

$$[x_3, x_4] = x_6 + x_7 - 3x_8$$

$$[x_3, x_5] = x_8$$

with $[x_i, x_j] = 0$, for $i < j$, otherwise. To verify characteristic nilpotence, it was not necessary to solve the system of derivation equations (in 64 variables) over Z. We found that $\dim(\mathrm{Der}(L_5))$ = 10. It sufficed then to exhibit 10 independent and strictly triangular derivations of L. One sees easily that any linear transformation $\tau: (x_1, x_2) \rightarrow (x_6, x_7, x_8)$ extends to a derivation, $\bar{\tau}$, of L with $\bar{\tau}(x_3) = [x_1, \tau(x_2)] - [x_2, \tau(x_1)]$ and $\bar{\tau}(L^3) = 0$ (Note that (x_6, x_7, x_8) is the transporter of L to Center(L)). The space of these is 6-dimensional and intersects the 7-dimensional space of inner derivations in a 3-dimensional subspace. Therefore $\dim(\mathrm{Der}(L)) = 10$ and L is characteristically nilpotent.

Note that, in one sense, the above 'typical' algebra does not have as 'few' outer derivations as the Dixmier-Lister example since the images of central derivations (i.e., derivations mapping to Center(L)) do not already span the outer derivation algebra. However, the Dixmier-Lister example has a derived algebra of co-dimension 4 and, as we already remarked, our random searches never produce such an example. This inspired a computer search amongst those algebras with $\dim(L) = 8$ and $\dim(L/L^2) = 4$; and this time the images of the central derivations did span the outer derivation algebras.

4.3 Characteristically Nilpotent Derived Algebras

Dixmier and Lister also showed that their example answered
in the negative the stronger question of whether every nilpotent
Lie algebra was a derived algebra of some Lie algebra. They then
asked whether a characteristically nilpotent Lie algebra can
ever be a derived algebra. With some effort we were able (see
[8]) to construct by hand an 18-dimensional algebra with a 16-
dimensional characteristically nilpotent derived algebra. However, it seemed natural recently to ask the computer the question.
Again generating random examples, we found

The algebras of dimension $\geq$ 13 *always had characteristically*
nilpotent derived algebras.

Here too, it was possible to find examples in characteristic 0.
In fact, the following algebra is an extension of the example in
Section 4.2. The algebra, L, has basis $\{x_1, x_2, \ldots, x_{13}\}$ and
multiplication

$$[x_1, x_2] = x_3 - x_4 + x_5 + x_7 - x_9 - x_{10} + 2x_{11} - x_{12} - x_{13}$$

$$[x_1, x_3] = x_4 + x_5 - x_6 - x_7 \quad - 2x_8 + 2x_9 + 2x_{13}$$

$$[x_1, x_4] = -x_5 - x_6 + x_7 - x_9 + x_{10} - x_{11} - x_{12} + x_{13}$$

$$[x_1, x_5] = x_6 + x_8 + x_9 + x_{10} + x_{11} - 2x_{12} + 2x_{13}$$

$$[x_1, x_6] = x_8 + 5x_9 - 2x_{10} - 4x_{11} + 54x_{12} - 2860x_{13}$$

$$[x_1, x_7] = -2x_8 - 2x_9 - x_{10} + x_{11} - 2x_{12}$$

$$[x_1, x_8] = x_9 + x_{11} + x_{12} + x_{13}$$

$$[x_1, x_9] = x_{11} + 4x_{12} + 194x_{13}$$

$$[x_1, x_{10}] = 2x_{11} + 5x_{12} + 10x_{13}$$

$$[x_1, x_{11}] = -5x_{12} - 5x_{13}$$

$$[x_1, x_{12}] = -15x_{13}$$

$$[x_2, x_3] = -x_5 - 2x_6 + x_8 + x_9 - x_{10} + x_{11} \quad -2x_{12} + x_{13}$$

$$[x_2, x_4] = -3x_6 - 2x_7 + x_8 + 2x_9 - 5x_{10} + 7x_{11} - 79x_{12} +$$

$$2441 \, x_{13}$$

$$[x_2, x_5] = x_6 + x_7 - x_9 + x_{11} - 2x_{13}$$

$$[x_2, x_6] = -2x_8 + 5x_9 - 4x_{10} + 4x_{11} + 213x_{12} - 9780x_{13}$$

$$[x_2, x_7] = 2x_8 + x_{10} + 2x_{11} - x_{12} - 2x_{13}$$

$$[x_2, x_8] = - x_{10} + x_{11} + x_{12} - 2x_{13}$$

$$[x_2, x_9] = -x_{11} - 20x_{12} - 160x_{13}$$

$$[x_2, x_{10}] = -40x_{12} + 435x_{13}$$

$$[x_2, x_{11}] = 175x_{13}$$

$$[x_3, x_4] = x_6 + x_7 - 3x_8 - 3x_9 + 2x_{10} + 6x_{11} + 149x_{12} -$$

$$2170x_{13}$$

$$[x_3, x_5] = x_8 + x_9 + x_{10} - 3x_{11} - 55x_{12} + 5176x_{13}$$

$$[x_3, x_6] = -2x_9 + x_{10} - 51x_{12} + 3372x_{13}$$

$$[x_3, x_7] = 2x_9 - 2x_{10} - 13x_{12} - 1064x_{13}$$

$$[x_3, x_8] = -x_{11} + 10x_{12} - 95x_{13}$$

$$[x_3, x_9] = 5x_{12} + 55x_{13}$$

$$[x_3, x_{10}] = 250x_{13}$$

$$[x_4, x_5] = 3x_9 - x_{10} + 5x_{11} + 76x_{12} - 2047x_{13}$$

$$[x_4, x_6] = x_{11} - 43x_{12} + 1847x_{13}$$

$$[x_4, x_7] = -x_{11} + 48x_{12} - 67x_{13}$$

$$[x_4, x_8] = -125x_{13}$$

$$[x_4, x_9] = -75x_{13}$$

$$[x_5, x_6] = 5x_{12} - 1290x_{13}$$

$$[x_5, x_7] = -20x_{12} + 950x_{13}$$

$$[x_5, x_8] = -75x_{13}$$

$$[x_6, x_7] = 150x_{13}$$

with $[x_i, x_j] = 0$ for $i < j$, otherwise. Again, making use of Proposition 2.1, we determined that $\dim(\text{Der}(L_7^2)) = 27$ and then managed to identify 27 independent, strictly triangular elements of $\text{Der}(L^2)$.

4.4   The Nilpotent Lie Algebras of Dimension 7

The computer never found a characteristically nilpotent Lie algebra of dimension 7 even though G. Favre [3] had shown that one exists. However, like Favre's example the generated algebras did have derivation algebras of dimension 10. One suspects then that there are "characteristically nilpotent points" within the open set of multiplications yielding minimal derivation algebras but such points do not themselves form an open set. Furthermore, this suggests that it ought to be possible to deform the characteristically nilpotent example to non-characteristically nilpotent ones. Indeed, we find the following family, L_t, of nilpotent Lie algebras on the 7-dimension space with basis $\{x_1, x_2, \ldots, x_7\}$: The multiplication in L_t is given by

$$[x_1, x_i] = x_{i+1} \qquad \text{for } 2 \leq i \leq 6$$

$$[x_3, x_4] = x_7$$

$$[x_3, x_2] = x_6 + tx_5$$

$$[x_4, x_2] = x_7 + tx_6$$

$$[x_5, x_2] = (t+1)\ x_7$$

$$[x_i, x_j] - 0 \qquad \text{for } i + j > 7.$$

The algebra L_0 is Favre's example which is characteristically nilpotent. However, in general, L_t has the derivation δ_t where

$$\delta_t(x_1) = 4t^2 x_1 + 2x_2$$

$$\delta_t(x_2) = 8t^2 x_2 + (6t+2)x_3 - (5t+1)x_4$$

$$\delta_t(x_3) = 12t^2 x_3 + (6t+2)x_4 - (5t+1)x_5$$

$$\delta_t(x_4) = 16t^2 x_4 + (4t+2)x_5 - (5t+3)x_6$$

$$\delta_t(x_5) = 20t^2 x_5 + (2t+2)x_6 - (5t+5)x_7$$

$$\delta_t(x_6) = 24t^2 x_6$$

$$\delta_t(x_7) = 28t^2 x_7.$$

Thus L_t is *not* characteristically nilpotent for $t \neq 0$.

Acknowledgment

The author is pleased to acknowledge research support from Research Corporation.

References

[1]   J. Dixmier, *Cohomologie des algèbres de Lie nilpotentes*,
       Acta Sci. Math. Szeged, 16(1955), 246-250.

[2]   J. Dixmier and W. Lister, *Derivations of nilpotent Lie
       algebras*, Proc. A.M.S., 8(1957), 155-158.

[3]   G. Favre, *Une algèbre de Lie caractéristiquement nilpotente
       de dimension 7*, C.R. Acad. Sc. Paris Ser A, 274(1972), 1338-
       1339.

[4]   M. Gauger, *On the classification of metabelian Lie algebras*,
       Trans. A.M.S., 179(1973), 293-328.

[5]   N. Jacobson, *A note on automorphisms and derivations of
       Lie algebras*, Proc. A.M.S., 6(1955), 281-283.

[6]   G. Leger and E. Luks, *On derivations and holomorphs of
       nilpotent Lie algebras*, Nagoya Math. J., 44(1971), 39-50.

[7]   _____, *Correction and comment concerning
       "On derivations and holomorphs of nilpotent Lie algebras,"*
       Nagoya Math. J., 59(1975), 217-218.

[8]   E. Luks, *A characteristically nilpotent Lie algebra can
       be a derived algebra*, Proc. A.M.S., 56(1976), 42-44.

[9]   V.V. Morosov, *Classification of nilpotent Lie algebras of
       sixth order*, Izv. Vyss Ucebn. Zaved. Mathematika, no.4(5),
       (1958), 161-171 (Russian).

[10]   A. Nijenhuis and R. Richardson, *Deformations of Lie algebra
       structures*, J. Math and Mech. 17(1967), 89-105.

[11]   E. Schenkman, *On the derivation algebra and the holomorph
       of a nilpotent algebra*, Mem. A.M.S., no. 14(1955), 15-22.

[12]   M. Vergne, *Cohomologie des algèbres de Lie nilpotentes.
       Application of l'etude de la variété des algèbres de Lie
       nilpotentes*, Bull. Soc. Math. France, 98(1970), 81-116.

Department of Mathematics
Bucknell University
Lewisburg, Pennsylvania 17837

INTEGER CLEBSCH-GORDAN COEFFICIENTS
FOR LIE ALGEBRA REPRESENTATIONS

Johan G. F. Belinfante

1. Introduction

The importance of the Lie theory for applied mathematics
and the physical sciences has grown substantially in recent
years. In part, the renewed interest in the Lie theory can
be attributed directly to the development of improved computational methods. Modern methods permit many important calculations involving the irreducible representations of semisimple
Lie algebras to be carried out in a systematic and uniform
way on an electronic digital computing machine using integer
mode arithmetic. We present here an historical survey of the
use of computers in Lie algebra theory, with particular reference to computing the coupling and recoupling coefficients for
the irreducible representations of simple Lie algebras of
arbitrary type using Chevalley bases.

In addition to the intrinsic mathematical interest in
the coupling and recoupling coefficients, an important motivation for computing these numbers comes from their extensive
applications in atomic, nuclear and elementary particle
physics. There already exists an extensive literature on
the problem of computing these coefficients, and progress
continues to be made. Yet, despite this impressive literature,
a complete resolution of the problem does not yet exist in
print. The coupling coefficients needed in applications can
often be computed using various special tricks, but no generally applicable algorithm for doing this has been published.

209

We shall have achieved our aim if, in presenting the following survey of the literature on this problem, we have enabled others to perceive at least the general outline for such a universal algorithm by which one may compute the coupling coefficients for any simple Lie algebra over the complex number field.

For convenience, we have arranged the material in chronological order, based on publication dates. In many cases, of course, the actual work was done a year or two earlier. We recognize that the bibliography is not complete, and we hereby offer our apologies to those whose work has been left out.

2.   Classical Work, Before 1960

E. B. Dynkin in 1947 published an exposition of Lie algebra theory starting from first principles [1]. This survey, based on the classical work of S. Lie, W. Killing, E. Cartan, H. Weyl and A. I. Mal'cev, contains important algorithms for Lie algebras still used today. A new feature in Dynkin's treatment was the introduction of schemas, now called Dynkin diagrams, which summarize information about the angles and relative lengths of a system of simple roots of a semisimple Lie algebra over the field of complex numbers.

Let us denote by ℓ the rank of a semisimple complex Lie algebra. By definition, the rank is the dimension of any Cartan subalgebra. The simple roots, α_1, α_2, $\ldots \alpha_\ell$ are certain nonzero linear forms on a Cartan subalgebra. For any nonzero linear form α on the Cartan subalgebra, we define another form $\hat{\alpha}$ by

$$\hat{\alpha} = \frac{2\alpha}{(\alpha,\alpha)}.$$

The Dynkin diagram determines an integer array M_{ij} which is equal to a numerical factor times (α_i, α_j). The unknown numerical factor drops out when we compute the integer Cartan matrix

$$A_{ij} = (\hat{\alpha}_i, \alpha_j) = 2 \, M_{ij}/M_{ii}.$$

The entire Lie algebra can be reconstructed from the Cartan matrix by classical algorithms explained in Dynkin's paper.

The irreducible finite-dimensional representations of certain semisimple Lie algebras were studied in 1950 by I. M. Gel'fand and M. L. Zetlin using explicit formulas based on chains of subalgebras [2,3]. To generalize this method to other Lie algebras, it is necessary to study the subalgebra structure of a given semisimple Lie algebra.

In 1952, such a study of subalgebras was initiated by E. B. Dynkin [4]. In the appendix to his paper on maximal subalgebras, Dynkin summarized the main facts of the theory of irreducible representations of semisimple Lie albegras over the complex number field. Each irreducible representation is characterized by its highest weight λ. Any weight μ is an integer linear combination of the highest weights $\lambda_1, \lambda_2, \ldots \lambda_\ell$ of certain basic representations,

$$\mu = m_1 \lambda_1 + m_2 \lambda_2 + \ldots + m_2 \lambda_2$$

These basic weights $\lambda_1, \lambda_2, \ldots \lambda_\ell$ are geometrically determined by the condition that $(\hat{\alpha}_i, \lambda_j) = \delta_{ij}$ is the Kronecker delta, equal to one if $i = j$ and zero if $i \neq j$. The integer coefficients

$$m_i = (\mu, \hat{\alpha}_i)$$

are called the Dynkin indices of the weight μ. It is convenient in computational work to express all roots and weights in terms of their Dynkin indices. In particular, the Cartan matrix represents the Dynkin indices of the simple roots. The

Dynkin indices of the highest weight of any representation are non-negative integers, and conversely, to any set of non-negative integers $(n_1, n_2, \ldots n_\ell)$ there corresponds, up to equivalence, a unique irreducible representation with highest weight $n_1 \lambda_1 + n_2 \lambda_2 + \ldots + n_\ell \lambda_\ell$. All information about an irreducible representation can be expressed in terms of the highest weight. In particular, there are two formulas due to H. Weyl which can be used to compute the dimension and the value of the second order Casimir operator.

In this same appendix, Dynkin describes an algorithm for computing the entire weight system of an irreducible representation from its highest weight. The weight system is built up layer by layer, each weight on a given layer being obtained from some weight on the preceding layer by subtracting some simple root. For the total number of layers, Dynkin gives a formula, which can be written as

$$1 + 2 \sum_{i,j} n_i \, (A^{-1})_{ij}.$$

Here the n_i are the Dynkin indices of the highest weight, while $(A^{-1})_{ij}$ are the entries of the inverse of the Cartan matrix. The inverse of the Cartan matrix can also be used to compute the metric tensor

$$g_{ij} = (\lambda_i, \lambda_j) = \frac{1}{2} (\alpha_i, \alpha_i)(A^{-1})_{ij}.$$

This metric tensor is needed to compute inner products of weights when we express them in terms of Dynkin indices. Note that since the Dynkin diagram only gives the relative lengths of the simple roots, the metric tensor is as yet known only up to an overall factor. We can determine this factor by using the fact that the Casimir operator has the value one in the adjoint representation.

Another paper by Dynkin, also written in 1952, deals with semisimple subalgebras [5]. In the introduction to this paper, explicit tables are given for the metric tensor $g_{ij} = (\lambda_i, \lambda_j)$ and its inverse $g^{ij} = (\hat{\alpha}_i, \hat{\alpha}_j)$.

A new algorithm for computing the characters of the irreducible representations of semisimple Lie algebras was developed by H. Freudenthal in 1954. This procedure is more efficient than an older Weyl formula using 'girdle division.' The essence of Freudenthal's method lies in the construction of his 'table D' and the use of an inductive formula for computing the multiplicities of the dominant weights on a given layer in terms of the multiplicities of the weights on previous layer [6]. By means of this procedure, the characters of even the exceptional simple Lie algebra E_8 could be obtained by hand. This algorithm remains today the quickest way to compute characters.

A valuable computational tool for Lie algebras has grown out of the discovery by C. Chevalley in 1955 of a new class of finite simple groups related to Lie algebras [7]. Chevalley found a systematic construction of bases for Lie algebras with respect to which all calculations can be carried out using integer-mode arithmetic. The structure constants are integers computed as follows. First, for each nonzero root α, a co-root h_α is defined. This co-root belongs to the Cartan subalgebra and satisfies

$$(h_\alpha, h) = \hat{\alpha}(h)$$

for all elements h in the Cartan subalgebra. Next, Chevalley shows that one can assign a root vector x_α to each non-zero root α such that

$$[h, x_\alpha] = \alpha(h)\, x$$
$$[x_\alpha, x_{-\alpha}] = h_\alpha$$

and

$$[x_\alpha, x_\beta] = N_{\alpha,\beta}\, x_{\alpha+\beta}$$

where the integers $N_{\alpha,\beta}$ satisfy $N_{-\alpha,-\beta} = -N_{\alpha,\beta}$. These integers are determined up to sign by

$$N_{\alpha,\beta} = \pm(p+1),$$

where $\beta-pa,\ldots,$ $\beta+q\alpha$ is the α-string through the root β.

To help dispel some of the possibly mysterious aspects surrounding the use of Chevalley bases, we consider the most elementary example, the simple Lie algebra A_1. This is the Lie algebra studied in the quantum theory of angular momentum. It is the complexification of the real Lie algebra of the Lie group $SU(2)$. Here one has only a single positive root α, and we may set $e = x_\alpha$, $f = x_{-\alpha}$ and $h = h_\alpha$. The Lie products are

$$[e,f] = h$$
$$[h,e] = 2e$$
$$[h,f] = -2f.$$

The connection with the notation used in the quantum theory of angular momentum is given by the formulas

$$h = 2j_3$$
$$e = j_1 + \sqrt{-1}\ j_2$$
$$f = j_1 - \sqrt{-1}\ j_2.$$

We call attention here to the factor 2 in the equation $h = 2j_3$ which serves to eliminate all the half-integers abounding in angular momentum theory.

R. Bivins, N. Metropolis, M. Rotenberg and J. K. Wooten, Jr. used electronic computers to prepare extensive tables of Clebsch-Gordan and Racah coefficients for $SU(2)$. These tables, published in 1959, are used by physicists and chemists in a wide variety of fields [8].

It was also in the year 1959 that B. Kostant published his elegant formula for the multiplicity of a weight, using a certain partition function [9]. This closed formula, like the

Weyl formula, involves summing over the Weyl group, but it
does not require division. The obvious question arises whether
this new formula is an improvement over the Freudenthal algorithm.

3. Work in the 1960's

R. Steinberg in 1961 used the Kostant formula to develop
a formula for the reduction of the tensor product of two irreducible modules over a semisimple Lie algebra as a direct sum of
irreducible submodules [10]. This expression for the multiplicities of the irreducible submodules occurring in the reduction involves the partition function and a double summation over
the Weyl group. The Steinberg formula reduces to the original
Clebsch-Gordan results when one specializes to the simple Lie
algebra A_1. The work of Steinberg of course concerns only the
Clebsch-Gordan series, not the Clebsch-Gordan coefficients.

In that same year, physicists, led by M. Gell-Mann, found
experimental evidence that the Lie group SU(3) is an approxi-
mate symmetry in the dynamics of the strongly interacting
elementary particles [11]. This work created widespread
interest in this particular Lie group as well as in the application of Lie group theory in general.

In the following year, N. Jacobson published the first
book on Lie algebras in the English language [12]. Many of
the above-mentioned algorithms for Lie algebraic calculations
are described in detail in this book.

Also in 1962 an interesting monograph by I. B. Levinson,
V. V. Vanagas and A. P. Yutsis appeared in which a graphical
calculus for handling Clebsch-Gordan and Racah coefficients
for SU(2) was developed [13]. This technique allows one to
draw pictures giving insight into the algebraic formulas.

Extensive hand calculations were done in 1963 by M. Konuma, K. Shima and M. Wada, and by other physicists, for the rank 2 and rank 3 simple Lie algebras needed in applications [14]. The needs of particle physics also led J. J. de Swart to compute the Clebsch-Gordan coefficients for SU(3) by hand [15].

By 1964 the computation of the coupling and recoupling coefficients for the group SU(2) had progressed to the point that representations of dimension up to a hundred could be handled. R. M. Baer and M.G. Redlich reported that such large integers occur in these calculations that it became necessary to introduce multiple precision fixed point arithmetic subroutines [16].

While studying extensions of Chevalley's method of constructing finite simple groups, R. Ree in 1964 introduced the concept of a Chevalley basis for a Lie module [17]. With respect to such a basis, the action of the Lie algebra on the module can be described by integer matrices. Ree first proved the existence of Chevalley bases for certain elementary irreducible representations for each type of simple Lie algebra, and then used Cartan composition to extend the result to arbitrary irreducible representations.

D. A. Smith gave a new proof of the existence of Chevalley bases for Lie modules in 1965 which avoids Cartan composition and the numerous case considerations required in Ree's proof. Instead, he extends Chevalley's formulas for the Lie algebra to its universal enveloping associative algebra, and uses the fact that every irreducible module is a cyclic module over the enveloping algebra [18].

The application of Chevalley bases to the construction of finite simple groups is the topic of a review article written by R. W. Carter in 1965. Carter covers not only the original work by Chevalley, but also the later extensions of the theory by R. Steinberg, J. L. Tits and R. Ree [19].

J. R. Derome and W. T. Sharp showed how to define the n-j coefficients for any compact group in 1965, and they derived a number of their algebraic properties [20]. The letter 'j' in the name 'n-j coefficient' stands for 'module.' Since all the n-j coefficients can be defined in terms of the 3-j coefficients, the basic computational problem is to calculate the latter.

While Derome and Sharp present no specific algorithm for obtaining the 3-j coefficients, it is not hard to describe such a procedure. The calculation of the 3-j symbols is mathematically equivalent to constructing a basis for the trivial submodule of the tensor product of three irreducible modules. The trivial submodule of a reducible module is by definition the set of all vectors in the module which are annihilated by every element of the Lie algebra. As we shall see later, the determination of the trivial submodule amounts to solving a system of homogeneous linear Diophantine equations.

The problem of finding integer solutions of homogeneous linear equations comes up over and over again in Lie algebraic problems. One can view the process of solving such equations as that of constructing a basis for a finitely generated abelian group subject to a finite number of relations [21]. A simple ALGOL procedure for doing this was written by D. A. Smith in 1966.

The proof of the existence of Chevalley bases for Lie modules was put into a particularly elegant form by B. Kostant in 1966. We briefly explain some of the ideas here. Consider a simple complex Lie algebra with N positive roots, $\alpha_1, \alpha_2, \ldots, \alpha_N$. For any sequence $S = (s_1, s_2, \ldots 2_N)$ of non-negative integers, define the 'raising' and 'lowering' elements,

$$e_s = \frac{(x_{\alpha_1})^{s_1}}{s_1!} \quad \frac{(x_{\alpha_2})^{s_2}}{s_2!} \quad \cdots \quad \frac{(x_{\alpha_N})^{s_N}}{s_N!}$$

$$f_s = \frac{(x_{-\alpha_1})^{s_1}}{s_1!} \quad \frac{(x_{-\alpha_2})^{s_2}}{s_2!} \quad \cdots \quad \frac{(x_{-\alpha_N})^{s_N}}{s_N!} ,$$

belonging to the universal enveloping algebra of the Lie algebra. If v is a nonzero vector of highest weight for an irreducible module, then the set of all the vectors $f_s v$ spans the
module. All but a finite number of these vectors are zero, and
the nonzero ones need not be linearly independent. From these
elements, however, one can extract a basis with respect to
which the module action can be expressed in terms of integer
matrices [22].

To illustrate these ideas, we turn again to the Lie algebra A_1. In the universal enveloping algebra, we introduce the
divided powers

$$e_k = \frac{e^k}{k!} , \qquad f_k = \frac{f^k}{k!} .$$

The highest weight of an irreducible module over A_1 is characterized by a single Dynkin index n, which can be $0, 1, 2, \cdots$.
The dimension of the module is $n+1$. If v is a nonzero vector
of highest weight, $hv = nv$, then the vectors $v, fv, f_2v, \cdots,$
$f_n v$ form a basis for the module. The matrices of the basis
vectors e, f, h of the Lie algebra A_1 with respect to this
basis for the module are as follows:

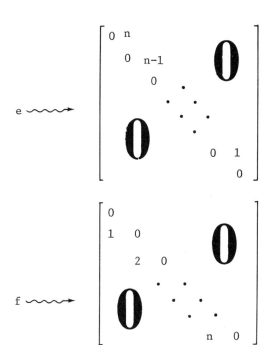

$$e \longrightarrow \begin{bmatrix} 0 & n & & & & & & \\ & 0 & n-1 & & & & \mathbf{0} & \\ & & 0 & \cdot & & & & \\ & & & \cdot & \cdot & & & \\ & & \cdot & & \cdot & & \\ \mathbf{0} & & & & \cdot & & \\ & & & & & 0 & 1 \\ & & & & & & 0 \end{bmatrix}$$

$$f \longrightarrow \begin{bmatrix} 0 & & & & & & \\ 1 & 0 & & & & \mathbf{0} & \\ & 2 & 0 & & & & \\ & & \cdot & \cdot & & & \\ \mathbf{0} & & \cdot & \cdot & & & \\ & & & \cdot & \cdot & & \\ & & & & n & 0 \end{bmatrix}$$

$$h \longrightarrow \begin{bmatrix} n & & & & & \\ & n-2 & & & \mathbf{0} & \\ & & \cdot & & & \\ & & & \cdot & & \\ \mathbf{0} & & & \cdot & \\ & & & & -n \end{bmatrix}$$

One should contrast these simple formulas for the action of the Lie algebra A_1 on a module with the square-root laden formulas traditionally employed in angular momentum theory. In the quantum theory of angular momentum, one writes

$$j = \frac{n}{2}, \qquad m = \frac{n}{2} - k,$$

and one uses the basis vectors

$$|j\ m\rangle = \sqrt{(j - m)!\ (j + m)!}\ f_{j-m}v.$$

From this one sees that all square roots in angular momentum theory can be removed simply by changing the normalization of the basis vectors.

In setting up the Chevalley basis for a simple Lie algebra, one needs to have a prescription for choosing the signs of the coefficients $N_{\alpha,\beta}$. These signs are to some extent arbitrary, but they must satisfy

$$N_{\beta,\alpha} = -N_{\alpha,\beta} = N_{-\alpha,-\beta}$$

and they must be chosen in a manner consistent with the Jacobi identity for the Lie algebra [23]. An algorithm for making such a consistent choice of signs in a Chevalley basis was described by J. L. Tits in 1966. To explain this algorithm, we need to introduce some terminology. Let us define an *addable pair* of roots (α,β) to be a pair of nonzero roots α and β such that $\alpha+\beta$ is also a nonzero root. An addable pair (α,β) is called a *special pair* if α and $\beta-\alpha$ are positive. A special pair (α,β) is *extraspecial* if every special pair (γ,δ) with $\alpha+\beta = \gamma + \delta$ satisfies $\alpha \leq \gamma$ and hence also $\delta \leq \beta$. The signs in a Chevalley basis can be chosen arbitrarily for the extraspecial pairs. The Jacobi identity is used to extend the sign prescription to the special pairs, and finally the sign for any addable pair is obtained from a special pair by the *triangle rule:* if $\alpha+\beta+\gamma=0$, then $N_{\alpha,\beta}$, $N_{\beta,\gamma}$ and $N_{\gamma,\alpha}$ all have the same sign.

In the mid-1960's physicists began to set up computer programs tackling individual Lie algebras on a piecemeal basis. Particular attention was directed to the Lie algebra of the group SU(3), partly because it is the next more complicated case after SU(2), and partly because it has important applications to the collective model of the atomic nucleus, as well as to elementary particle physics. For example, M. Herttua and P. Jauho in 1966 reported on an ALGOL program to compute the Clebsch-Gordan series for SU(3). In their work, the Clebsch-Gordan coefficients were not discussed [24].

While setting up a computer program for SU(2) Racah coefficients, J. Stein in 1967 discovered a binary algorithm for obtaining the greatest common divisor of two integers [25]. Any greatest common divisor algorithm can be used to speed up the solution of linear Diophantine equations. The binary algorithm finds the greatest common divisor using shifting, parity testing, and subtraction. The algorithm is fast because, unlike Euclid's algorithm, it requires no divisions. D. E. Knuth *(The Art of Computer Programming,* vol. 2, section 4.5.2) says that this algorithm was actually discovered earlier by R. L. Silver and J. Terzian.

J. L. Tits in 1967 published tables providing some basic information concerning simple Lie groups and their representations [26].

L. C. Biedenharn and various collaborators embarked on an ambitious program to compute the Clebsch-Gordan and Racah coefficients for the unitary groups [27]. They did not use Chevalley bases, but instead used the subgroup chain U(n) ⊃ U(n-1) ⊃ ... ⊃ U(2), following the 1950 Gelfand-Zetlin paper. In 1967 they proposed a canonical definition for the coupling coefficients of the unitary groups, based on an embedding of the irreducible representations of U(n) into the totally symmetric irreducible representations of $U(n^2)$.

In the following year, K. B. Wolf announced a set of
FORTRAN subroutines for handling polynomial bases for representations of the group $U(n)$, using the same subgroup-chain
strategy [28].

V. K. Agrawala and J. G. F. Belinfante in 1968 developed
a graphical recoupling theory for compact groups, based on the
algebraic formalism of J. R. Derome and W. T. Sharp. Lines
are used to represent modules, and nodes represent module
homomorphisms. Parallel lines represent tensor products,
while crossed lines represent exchange operators [29].   For
example, the graph

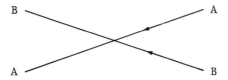

represents the operator taking the vector $u \otimes v$ in the module
$A \otimes B$ into the vector $v \otimes u$ in the module $B \otimes A$.

Using this notation, one can distinguish two kinds of
3-j symbols:

Here, the solid lines represent irreducible modules A, B, and C,
while the dotted lines represent the trivial submodule of their
tensor product. The graph on the left represents the inclusion
mapping, while the graph on the right represents the module
homomorphism which projects the tensor product module into its
trivial submodule. This is unique because the trivial submodule
is an isotypical component. This projection can be constructed

explicitly by averaging over the group, using the Haar integral.

All other n-j symbols can be defined in terms of the 3-j symbols. For example, the 9-j symbol is defined by the following picture.

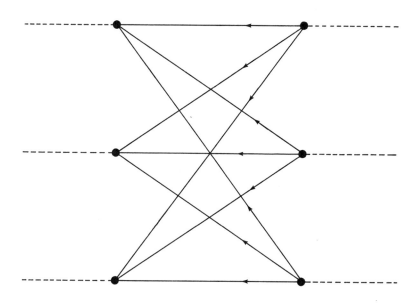

The calculation of 3-j symbols using Chevalley bases and Diophantine linear equations was illustrated for the simple Lie algebra A_2 in 1969 by J. G. F. Belinfante and B. Kolman in the third of a series of survey articles on Lie algebras and their representations [30].

To explain this method, we consider here a simpler problem, the computation of the 3-j coefficients for spinor-spinor-vector coupling in the simple Lie algebra A_1. The two spinor modules both have Dynkin index n = 1 (spin $j = \frac{1}{2}$), while the vector module has Dynkin index n = 2 (spin j = 1). If v and v' are highest weight vectors for the spinor modules, and v" a highest weight vector for the vector module, then a basis for the tensor product of these three irreducible modules is given by the vectors

$$v_{abc} = f_a v \otimes f_b v' \otimes f_c v''$$

where a, b = 0, 1 and c = 0, 1, 2. As before, the elements $f_n = f^n/n!$ are divided powers in the universal enveloping algebra.

The problem is to find the trivial submodule, consisting of vectors

$$t = \sum_{a,b,c} t_{abc} \, v_{abc}$$

satisfying et = ft = ht = 0. The numerical coefficients t_{abc} are the 3-j coefficients. The equation ht = 0 implies t_{abc} = 0 unless a + b + c = 2. The equation ft = 0 yields a system of homogeneous equations,

$$0 = t_{002} + 2 \, t_{101}$$
$$0 = t_{002} + 2 \, t_{011}$$
$$0 = t_{101} + t_{011} + t_{110}.$$

The remaining condition et = 0 gives nothing new. The solution of the Diophantine system is

$$t_{110} = -2 \, t_{101} = -2 \, t_{011} = t_{002},$$

and therefore

$$2 \, v_{110} - v_{101} - v_{011} + 2 \, v_{002}$$

is a basis for the trivial submodule. We see that the trivial submodule in this case happens to be one-dimensional. This holds generally for the Lie algebra A_1, but not for other simple Lie algebras.

Extensive calculations of the characters and related basic information for irreducible representations of simple Lie algebras were begun in 1968 by V. K. Agrawala and J. G. F. Belinfante on a UNIVAC 1108 machine using FORTRAN V. Up to that time, such computations reported in the leterature had been done by hand and were mostly limited to algebras of rank three or less. The program uses the Dynkin layer algorithm to obtain weight systems, and the Freudenthal formula for weight multiplicities, making no use of Weyl reflections. The FORTRAN program has subroutines to automatically scan through the simple Lie algebras of all types with rank up to eight, and to examine those irreducible modules found to have dimension less than a thousand. The dimensions of the modules and the values of the second order Casimir operator are found from Weyl's formula. There are also subroutines to print out useful information about the duals of the modules, and the Young tableaux corresponding to Dynkin indices. Originally, an integer-mode version of the Gauss-Jordan reduction algorithm was used to invert the Cartan matrices, but this caused overflow problems for the C-type algebras of ranks 7 and 8, and for the D-type algebras of ranks 6, 7 and 8. No overflow occurs for A-type algebras up to rank 20. Perhaps this overflow problem could be eliminated by renumbering the simple roots. In the shortened ALGOL version of these programs published in 1969, the problems associated with the Cartan matrix inversion were bypassed by using empirical regularities found in Dynkin's tables [31].

4. Recent Work, Since 1970

A question raised in the paper by Agrawala and Belinfante is whether the Freudenthal algorithm is actually the most efficient method for computing characters. For simplicity they had used the Freudenthal formula to find the multiplicity of every weight, disregarding simplifications possible by using Weyl reflections . M. I. Krusemeyer in 1971 published an improved ALGOL 60 procedure in which, by analogy with most hand computations, the Freudenthal formula is used to compute only the multiplicities of the dominant weights. Other weights encountered in the course of the calculation are transformed into dominant weights by Weyl reflections [32].

Besides the Freudenthal algorithm for computing characters
of Lie modules, there are several alternative methods which
involve the Weyl group. R. E. Beck and B. Kolman in 1971 constructed a program to generate the Weyl group on a computer, paving
the way for a comparison of the various algorithms [33]. Their
procedure is to first generate a subgroup isomorphic to a symmetric group S_n, and then to generate the Weyl group by coset enum-
eration, using an explicit presentation of the Weyl group. Each
element of the Weyl group is stored as a word, using the simple
Weyl reflections as alphabet. The large size of the Weyl group
causes storage problems, limiting the methods to rank four. More
compact storage is possible by breaking each word into syllables
and storing only the syllables [40].

Consider for example the Weyl group of the simple Lie algebra A_3. Let 0 denote the identity element, and let 1, 2 and 3
denote the Weyl reflections corresponding to the first, second
and third simple roots, respectively. The Weyl group has 24
elements, and the longest word is 121321, which has six letters.
So one could require 6 × 24 = 144 storage locations to store the
words directly as an array. Each word however can be constructed
out of three syllables. The first syllable is either 0 or 1, the
second is 0, 2 or 21, and the third is 0, 3, 32 or 321. The
storage needed to contain these syllables as an array is only
3×4×3 = 36 locations. For higher rank Lie algebras, the savings
are even better.

It is well-known that by employing higher-order and polarized
Casimir invariants, one can compute 6-j symbols directly without
first setting up the 3-j symbols. V. K. Agrawala and J. G. F.
Belinfante in 1971 published ALGOL procedures for the relevant
Casimir invariants for SU(n) representations, obtained via their
diagram calculus [34].

N. Burgoyne in 1971 reported the first successful computer

implementation of algorithms involving Chevalley bases for Lie
modules [35]. For each weight μ of an irreducible module with
highest weight λ he defines a matrix whose entries are integers
a_{ST} defined by

$$e_S f_T v = a_{ST} v$$

Here, as before, $S = (s_1, \ldots, s_N)$ and $T = (t_1, t_2, \ldots, t_N)$ are
sequences of non-negative integers satisfying

$$\sum_{k=1}^{N} s_k \alpha_k = \sum_{k=1}^{N} t_k \alpha_k = \lambda - \mu,$$

and v is a nonzero vector of highest weight: $hv = \lambda(h) v$. The
matrices a_{ST} can be computed directly from the commutation
relations of the Lie algebra. To compute a Chevalley basis for
a Lie module, one needs to know all linear relations satisfied
by the vectors $f_T v$. If

$$\sum_T r_T f_T v = 0$$

is such a relation, then applying e_S, we obtain

$$\sum_T a_{ST} r_T = 0.$$

So, all these relations can be computed by finding the null
space of the matrix a_{ST}.

In another paper, N. Burgoyne and C. Williamson computed
the elementary divisors of the matrices a_{ST} to obtain results
on the multiplicites of weights in the characteristic p case [36].
They remark that their algorithm for setting up the matrix a_{ST}
has a long execution time. These programs were written in an
assembly language for an IBM 360-50 machine.

To speed up the solution of systems of Diophantine linear
equations, one needs not only to be able to calculate the greatest
common divisor d of a list of integers $a_1, a_2, \ldots a_n$, but one also
needs to be able to produce a set of multipliers $x_1, x_2, \ldots, x_n$ such
that $d = a_1 x_1 + a_2 x_2 + \ldots + a_n x_n$. G. H. Bradley in 1972 published

an ALGOL procedure for this, which is based on a modification of Euclid's algorithm [37]. One can gain a factor two in efficiency by allowing negative remainders in Euclid's algorithm.

D. N. Verma in 1971 used a computer to study a conjecture concerning a certain harmonic polynomial related to Weyl's dimension formula [38]. This conjecture was settled in 1974 by S. G. Hulsurkar [51].

R. E. Beck and B. Kolman in 1972 published a survey of their further experience using computers to study representations of Lie algebras [39]. They also reported on yet another variant of Freudenthal's algorithm in which one again computes only the multiplicities of the dominant weights, as in Krusemeyer's version [41]. The difference is now that one does not keep transforming every weight into a dominant weight, but instead one runs through the entire weight system just once, applying only a single Weyl reflection on each of the non-dominant weights, transforming them into higher weights with known multiplicities [47].

During this same period, J. G. F. Belinfante and B. Kolman published a monograph surveying the present status of the applied theory of Lie groups, Lie algebras, and their representations. This monograph concentrates on presenting the results needed to understand the current applied literature, and it includes an account of some experience using computers [42].

Readable accounts of Chevalley bases can be found in the recent book on Lie algebras by J. E. Humphreys and in the book on Chevalley groups by R. W. Carter, both published in 1972 [43, 44].

R. E. Beck and B. Kolman completed their comparison of
algorithms for inner and outer multiplicites in 1973. They found
that for inner multiplicities, the Freudenthal algorithm is best
[45]. Methods requiring the generation of the full Weyl group
are less efficient because the Weyl group grows rapidly in size
with increasing rank. For outer multiplicities (Clebsch-Gordan
series), they report that a formula of G. Racah provides the
shortest computation times [46]. Although Racah's formula
appears to involve the Weyl group, one need not actually compute
the full Weyl group because many terms in Racah's formula usually
drop out for low-dimensional representations. By using only some
properties of the Weyl group, and an appropriate stopping condition, computer programs using Racah's formula are quite fast
[48,50].

Any module over a Lie algebra can by restriction also be
considered as a module over any subalgebra. An irreducible
module over a simple Lie algebra will generally be reducible
upon restriction to a semisimple subalgebra, and its decomposition into irreducible submodules over the subalgebra is known as
a branching rule. Branching rules are important when one uses
the subgroup-chain approach to the coupling coefficients. Extensive computer-generated tables of branching rules were published
by J. Patera and D. Sankoff in 1973. This work also includes a
list of all irreducible modules of dimension less than a thousand over any simple Lie algebra of rank up to eight [49].

Many ideas used in the algorithms we have been discussing
first arose in the study of Chevalley's finite simple groups.
As one might expect, the theory of Chevalley groups has itself
continued to be developed over the past twenty years. We mention here just one recent development in this field, the publication by H. Behr in 1975 of explicit presentations for these
groups [52].

5. Outlook

We have surveyed the literature on algorithms concerning the representations of simple Lie algebras, with special reference to the problem of computing the 3-j coefficients using Chevalley bases. Since many of the relevant algorithms were originally developed in connection with hand calculations, only informal verbal descriptions of them were given in the literature. The availability of computing machines has led to renewed interest in the development of uniform methods for studying representations of Lie groups. For computer use, the classical algorithms must first be translated into precise and explicit programs. Publication of such formal computer programs is necessary so they can be verified, and so that meaningful comparisons of running times can be made when there are competing algorithms which accomplish the same task.

Bibliography (arranged in chronological order)

[1]   E.B. Dynkin, *The structure of semisimple algebras*, Uspehi Nat. Nauk (N.S.), $\underline{2}$ (1947), 59-127.

[2]   I. M. Gel'fand and M. L. Zetlin, *Finite-dimensional representations of the groups of unimodular matrices*, Doklady Akad. Nauk SSSR, $\underline{71}$,(1950), 825-828.

[3]   I. M. Gel'fand and M. L. Zetlin, *Finite-dimensional representations of groups of orthogonal matrices*, Doklady Akad. Nauk SSR, $\underline{71}$(1950), 1017-1020.

[4]   E. B. Dynkin, *The maximal subgroups of the classical groups*, Trudy Mosk. Mat. Obšč., $\underline{1}$ (1952), 39-166.

[5]   E. B. Dynkin, *Semisimple subalgebras of semisimple Lie algebras*, Mat. Sbornik, $\underline{30}$ (1952), 349-462.

[6]   H. Freudenthal, *On the calculation of the characters of semisimple Lie groups*, *II*, Indag. Math, $\underline{16}$(1954), 487-491.

[7] C. Chevalley, *On certain simple groups*, Tôhoku Math. J.(2)
     7,(1955), 14-66.

[8] R. Bivins, N. Metropolis, M. Rotenberg and J.K. Wooten, Jr.
     The 3-j and 6-j Symbols, M.I.T. Press, Cambridge, Mass.
     (1959).

[9] B. Kostant, *A formula for the multiplicity of a weight*,
     Trans. Amer. Math. Soc., 93(1959), 53-73

[10] R. Steinberg, *A general Clebsch-Gordan theorem*, Bull. Amer.
     Math. Soc., 67(1961), 401-407.

[11] M. Gell-Mann, *The eightfold way: a theory of strong
     interaction symmetry*, California Inst. of Technology
     Lab. Report CTSL-20, (1961).

[12] N. Jacobson, *Lie Algebras*, Interscience, J. Wiley & Sons,
     New York (1962).

[13] I. B. Levinson, V.V. Vanagas and A. P. Yutsis, *Mathemati-
     cal Apparatus of the Theory of Angular Momentum*, Israel
     Program for Scientific Translations, S. Monson, Jerusalem,
     Israel.(1962)

[14] M. Konuma, K. Shima and M. Wada, *Simple Lie algebras at
     rank 3 and symmetries of elementary particles in the
     strong interactions*, Progress in Theoretical Physics
     (Japan) Supplement, 28(1963), 1-128.

[15] J. J. de Swart, *The octet model and its Clebsch-Gordan
     coefficients*, Revs. Modern Physics, 35(1963), 916-939.

[16] R. M. Baer and M. G. Redlich, *Multiple precision arith-
     metic and the exact calculation of the 3-j, 6-j and 9-j
     symbols*, Commun. A.C.M., 7(1964), 657-659.

[17] R. Ree, *Construction of certain semi-simple groups*,
     Canad. J. Math, 16(1964) 490-508.

[18] D.A. Smith, *Chevalley bases for Lie modules*, Trans.
     Amer. Math. Soc., 115(1965), 283-299.

[19] R. W. Carter, *Simple groups and simple Lie algebras*,
     J. London Math. Soc., 40(1965), 193-240.

[20] J. R. Derome and W. T. Sharp, *Racah algebra for an
     arbitrary group*, 6(1965), 1584-1590.

[21]   D. A. Smith, *A basis algorithm for finitely-generated abelian groups*, Mathematical Algorithms, $\underline{1}$(1966), 13-35.

[22]   B. Kostant, *Groups over Z*, in *Algebraic Groups and Discontinuous Subgroups*, Proc. Symp. Pure Math., American Math. Soc., Providence, R.I., $\underline{9}$(1966), 90-98.

[23]   J. L. Tits, *On the structure constants and the existence theorem for semi-simple Lie algebras*, Inst. Hautes Études Sci. Publ. Math, $\underline{31}$(1966), 21-58.

[24]   M. Herttua and P. Jauho, *A computer program for computation of irreducible representations of SU(3)*, Suomalaisen Tiedeakatemian Toimituksia (Annales Academiae Scientiarum Fennicae), Helsinki, Ser. A. VI., Physica $\underline{208}$(1966), 6pp.

[25]   J. Stein, *Computational problems associated with Racah algebra*, J. Comput. Phys., $\underline{1}$(1967), 397-405.

[26]   J. L. Tits, *Tables for the Simple Lie Groups and Their Representations*, Springer-Verlag, Berlin, (1967).

[27]   L. C. Biedenharn, A. Giovannini and J. D. Louck, *Canonical definition of Wigner coefficients in U_n*, J. Math. Phys., $\underline{8}$(1967), 691-700.

[28]   K. B. Wolf, *A set of FORTRAN subroutines for handling bases of group representations*, J. Comput. Phys., $\underline{2}$ (1968), 334-335.

[29]   V. K. Agrawala and J. G. F. Belinfante, *Graphical formulation of recoupling theory for any compact group*, Ann. of Physics, $\underline{49}$(1968), 130-170.

[30]   J. G. F. Belinfante and B. Kolman, *An introduction to Lie groups and Lie algebras, with applications. III. Computational methods and applications of representation theory*, SIAM Review, $\underline{11}$(1969), 510-543.

[31]   V. K. Agrawala and J. G. F. Belinfante, *Weight diagrams for Lie group representations: A computer implementation of Freudenthal's algorithm in ALGOL and FORTRAN*, Nordisk Tidskrift for Informationsbehandling (Sweden), BIT $\underline{9}$(1969), 301-314, [Erratum: to be published].

[32]   M. I. Krusemeyer, *Determining multiplicities of dominant weights in irreducible Lie algebra representations using a computer*, Nordisk Tidskrift for Informationsbehandling (Sweden) BIT $\underline{11}$(1971), 310-316.

[33]   R. E. Beck and B. Kolman, *Generation of the Weyl group on a computer*, J. Comput. Phys., 7(1971), 346-353.

[34]   V. K. Agrawala and J. G. F. Belinfante, *An algorithm for computing SU(n) invariants*, Nordisk Tidskrift for Informationsbehandling (Sweden), BIT 11(1971), 1-15.

[35]   N. Burgoyne, *Representation theory of finite groups and related topics*, Proc. Symp. Pure Math. AMS, 21(1971), 13-17.

[36]   N. Burgoyne and C. Williamson, *Some computations involving simple Lie algebras*, Proc. SIGSAM 2 ACM, (1971), 162-171.

[37]   G. H. Bradley, *Algorithms for Hermite and Smith normal matrices and linear Diophantine equations*, Math. of Comput., 25(1971), 897-907.

[38]   D. N. Verma, *Lie Groups and Their Representations: Proceedings of the Summer School of the Bolyai János Mathematical Society*, Akadémiai Kiadó. Budapest, Hungary, (1971), edited by I. M. Gelfand, (J. Wiley & Sons, New York, 1975), 653-705, *Role of affine Weyl groups in the representation theory of algebraic Chevalley groups and their Lie algebras*.

[39]   R. E. Beck and B. Kolman, *Computer approaches to the representations of Lie algebras*, J. Assoc. for Comput. Mach., 19(1972), 577-589.

[40]   R. E. Beck and B. Kolman, *Computer generated Weyl groups*, Computer Physics Communications, 3(1972), 155-158.

[41]   R. E. Beck and B. Kolman, *A computer implementation of Freudenthal's multiplicity formula*, Indag. Math., 34, (1972), 350-352.

[42]   J. G. F. Belinfante and B. Kolman, *A Survey of Lie Groups and Lie Algebras, with Computational Methods and Applications*, SIAM, Philadelphia, 1972.

[43]   J. E. Humphreys, *Introduction to Lie Algebras and Representation Theory*, Springer-Verlag, New York, 1972.

[44]   R. W. Carter, *Simple Groups of Lie Type*, Wiley-Interscience, London, 1972 .

[45]   R. E. Beck and B. Kolman, *Computers in Lie algebras. I. Calculation of inner multiplicities*, SIAM J. Appl. Math., 25(1973), 300-312.

[46]   R. E. Beck and B. Kolman, *Computers in Lie algebras. II. Calculation of outer multiplicities*, SIAM J. Appl. Math., 25(1973), 313-323.

[47]   R. E. Beck and B. Kolman, *Freudenthal's inner multiplicity formula*, Computer Phys. Comm., 6(1973), 24-29.

[48]   R. E. Beck and B. Kolman, *A stopping method for Racah's formulas*, J. Comput. Phys., 13(1973), 161-163.

[49]   J. Patera and D. Sankoff, *Tables of Branching Rules for Representation of Simple Lie Algebras*, Univ. Montrèal Press, 1973.

[50]   R. E. Beck and B. Kolman, *Racah's outer multiplicity formula*, Computer Phys. Comm., 8(1974), 95-100.

[51]   S. G. Hulsurkar, *Proof of Verma's conjecture on Weyl's dimension polynomial*, Inventiones Math., 27(1974), 45-52.

[52]   H. Behr, *Explicit presentation of Chevalley groups over Z*, Math. Zeits, 141(1975), 235-241.

School of Mathematics
Georgia Institute of Technology
Atlanta, Georgia   30332

THE COMPUTATION OF BRANCHING RULES FOR REPRESENTATIONS
OF SEMISIMPLE LIE ALGEBRAS

W. McKay, J. Patera and D. Sankoff

1. Introduction

Given a semisimple Lie algebra Z over the complex field and
its irreducible representation $\phi(Z)$ of finite dimension, when
Z is restricted to a semisimple subalgebra H, the representation
$\phi(Z)$ becomes a representation $\phi(H)$ which, in general, is reducible.
The splitting of the original representation $\phi(Z)$ into the direct
sum $\phi(H)$ of irreducible representations of H is called the
branching rule (further BR).

Applications of the semisimple complex Lie groups and/or
algebras have been proliferating and diversifying in mathematics
[2,4] and physics [8]. The knowledge of the branching rules for
relevant group-subgroup pairs and representations turned out to
be indispensable in applications. Many authors devoted their
interests to various aspects of this problem ([7] and references
therein).

The purpose of this paper is to present a method for the
BR computation, applicable to any semisimple Z and H $\subseteq$ Z. The
procedure based on [10] was implemented in the original version
of our program and the results were published in [11], where
the BR for all representations of dimension < 1000 of all simple
Lie algebras of rank $\leq$ 8 were reduced with respect to all maximal
semisimple subalgebras. The present improved version of our
computer program was most recently used to calculate the branching
rules for the 3875-dimensional representation of the exceptional
simple Lie algebra E_8 [9].

235

The distinct feature of our method is that, in principle, it does not impose any restriction on the types of semisimple Z, H or the representations of Z to which it can be applied. Naturally, some practical limits are needed for the computer calculations.

The computational procedure we adopt has three components. The first one which we repeatedly use is the generation of a weight system of an irreducible representation starting from the highest weight. Although the well-known algorithm has been programmed previously [1] our implementation was done independently. The second part of our procedure is the projection of a given weight system of a representation of an algebra into a weight system of a representation of the subalgebra which is followed by the third step: the selection of the highest weights from the projected system.

The method is described in Section 2 and the computer program is presented in Section 3. Section 4 contains several examples involving the BR for the algebra E_7 which is currently of particular interest in elementary particle physics [6].

2.   Method

a.   Notations

A semisimple Lie algebra Z decomposes into a product of simple ideals Z_i; we write

$$(1) \qquad\qquad Z = Z_1 * Z_2 * \ldots * Z_k,$$

where each Z_i is one of the following nine types of simple Lie algebras: the four classical series A_n, B_n, C_n, D_n (n=1, 2,...) and five exceptional algebras E_6, E_7, E_8, F_4, and G_2. Because of the isomorphisms

$$A_1 \simeq B_1 \simeq C_1 \simeq D_1, \quad C_2 \simeq B_2, \quad D_2 \simeq A_1 * A_1, \qquad D_3 \simeq A_3$$

we do not consider explicitly algebras B_1, C_1, D_1, B_2, D_2, or D_3.

Each semisimple Lie algebra of rank n is completely char-
acterized by its system of simple roots α_1, α_2, ..., α_n. The
roots are vectors of a real Euclidean space, their relative
length and pairwise mutual angles are fixed for algebras of
each type. We fix as well the length of the roots by putting

(2) $$(\alpha_{max}, \alpha_{max}) = 2,$$

where α_{max} is the longest of simple roots of the algebra. The
properties of simple roots are usually summarized in diagrams
as in Table 1. Thus each o denotes a root of length $\sqrt{2}$, the
black dots ● denote roots of lengths 1 in algebras of types
B_n, C_n, and F_4, while in G_2 the length of the shorter root is
$\sqrt{\frac{2}{3}}$. Roots not connected directly by any line are mutually
orthogonal, the scalar product of two simple roots connected by
lines equals −1.

For our purposes it is convenient to use the properties of
simple roots in a form of the modified Cartan matrix

(3) $$g^{ik} = \frac{2(\alpha_i, \alpha_k)}{(\alpha_k, \alpha_k)} \quad (i,k=1,2,\ldots,n),$$

where the simple roots are numbered as in Table 1. If an
algebra is not simple, its Cartan matrix (3) has a block diagonal form, each block corresponding to an ideal Z_i.

An irreducible representation of a semisimple Lie algebra
Z of rank n is denoted by the contravariant coordinates

(4) $$a_i = \frac{2(\Lambda, \alpha_i)}{(\alpha_i, \alpha_i)}$$

of its highest weight $\Lambda \equiv (a_1 a_2 \ldots a_n)$. The coordinates a_i of Λ are non-negative integers.

b.   Computation of the Weight System

Our procedure is an implementation of the algorithm [10]. Each weight of a representation is denoted by the n-tuple of its contravariant coordinates. These coordinates are always integers. Starting from the highest weight $(a_1 a_2 \ldots a_n)$ we subtract from it simple roots

(5) $$\alpha_i = (g^{i1} g^{i2} \ldots g^{in}),$$

where g^{ik} is given by (3).

If the algebra Z consists of more than one simple ideal, say $Z = Z_1 * Z_2$ with respective ranks n_1 and n_2, then the n_1 first coordinates refer to a representation of Z_1 and the remaining n_2 to Z_2.

Together with each weight we keep track of two numbers associated with it: the level of the weight system to which the weight belongs and its multiplicity.

Two weights belong to the same level if they are obtained by subtracting the same number of simple roots from the highest weight; the actual roots which are subtracted do not need to be the same. The number of levels L in the weight system of an representation $(a_1 a_2 \ldots a_n)$ equals

(6) $$L = 1+T = 1 + \sum_{i=1}^{n} r_i a_i,$$

where $(r_1 r_2 \ldots r_n)$ is the level vector [5]. These vectors depend on the algebra Z. For each simple Lie algebra they are shown in Table II. Thus the value of L or T, the number of gaps, characterizes the highest level of the weight system. Similarly (6) is sometimes used to determine the level to which any weight of the system belongs.

Our procedure is set up in such a way that we need to calculate only the restricted weight system W_Λ instead of all the
weights.   The W_Λ consists of the weights from the upper

(7)
$$\left[\frac{T+1}{2}\right] = \left[\frac{T_1 + T_2 + \ldots + T_k + 1}{2}\right]$$

levels, where [x] denotes integer part of x.   If T is even then
the lowest level L_ℓ belonging to W_Λ is the middle level of the
whole weight system,   if T is odd the representation does not
have a middle level.

The second number which goes with every weight is its
multiplicity.   The multiplicity n_M of a weight M is calculated
from multiplicities of weights on levels above that of M using
Freudenthal's recursion formula,

(8)
$$\{(\Lambda+\delta,\Lambda+\delta)-(M+\delta,M+\delta)\}n_M = 2 \sum_k \sum_{i=1}^{n} n_{M+k\alpha_i} (M+\alpha_i,\alpha_i).$$

Here Λ is the highest weight and δ is the vector which has all
contravariant coordinates (4) equal to 1 for any algebra, α_i is
a positive root.   The summation over k extends through all the
values k=1,2,..., for which $M+k\alpha_i$ belongs to the weight system
of Λ.

The scalar product of two vectors $P = (p_1 p_2 \cdots p_n)$ and
$Q = (q_1 q_2 \cdots q_n)$ of the weight space given in terms of their
coordinates, is calculated as

(9)
$$(P,Q) = \sum_{i,j=1}^{n} p_i g_{ij} q_j,$$

where the scalar product matrices (g_{ij}) are given in Table III
for each simple Z.

For checking the computations we need to know also the dimension $N(\Lambda)$ and the 2nd order index [12] $I(\Lambda)$ of each representation Λ. These quantities are calculated, respectively, as the sum of 0th and 2nd powers of lengths of the weights of the whole weight system, each weight being counted as many times as its multiplicity in Λ.

c.   Projection of Weight Systems

Given a representation $\Lambda \equiv (a_1 a_2 \cdots a_n)$ of a semisimple Z, we first calculate its restricted weight system W_Λ. In order to find the representation of a semisimple H, H $\subset$ Z, we project each weight $M \in W_\Lambda$ by means of a projection matrix F onto the subspace spanned by the simple roots of H. As a result we obtain the restricted weight system $W_{\Lambda'}$ of a (reducible) representation Λ' of H contained in Λ. Here we describe a practical way to construct the projection matrix F. For all maximal semisimple Lie algebras of simple Lie algebras of rank $\leq$ 8 the projection matrices F are shown in Table IV.

An embedding of H into Z is usually specified by the representation ω_f of H contained in the lowest dimensional representation ω of Z. The representations ω are shown in Table V.

In order to set up the projection one needs to order the weight system of ω_f according to the following two rules:

(a)   All weights are numbered according to levels; weights from a higher level are assigned smaller values of indices

(b)   Weights from the same level are numbered using some lexicographical ordering; the same ordering must be used on every level.

As a result we end up with an ordered weight system $M_1, M_2, \ldots,$ of ω_f, where M_1 is the highest weight. In practice it is not necessary to calculate more than n weights M_i. The projection of the vectors

$$\nu_1 = (10 \ldots 0)$$

$$\nu_2 = (010 \ldots 0)$$

$$\vdots \qquad \vdots$$

$$\nu_n = (0 \ldots 01)$$

of the weight space of ω is then found from Table VI in terms
of weights M_i of ω_f.

We notice that for algebras of types A_n, B_n, C_n, E_7, E_8,
F_4, and G_2 it suffices to use the first n weights of ω_f. For E_6
it is convenient to use the first 4 weights M_1, M_2, M_3, M_4 and
the two lowest ones M_{26} and M_{27}.

We thus construct a rectangular n'×n matrix F, where n' is
the rank of the subalgebra H, which can be applied to a weight
system of any representation of Z in order to project it onto
the n'-dimensional space of weights of representations of H.
Since the lexicographical ordering can be chosen in many different ways, the matrix F can have many forms all equivalent for
our purposes.

d.   Separation of Irreducible Components

Suppose that the weights of W_Λ of a given representation
$\Lambda \equiv (a_1 a_2 \ldots a_n)$ of Z have been projected by means of F into
the weights of a reducible representation Λ' of H. Our present
task is to describe how to select among the projected weights
those which are the highest weights Λ'_1, Λ'_2, $\ldots$ of irreducible
components of Λ'.

The projection matrix F is constructed in such a way that
W_Λ is projected into $W_{\Lambda'}$, with the possible exception of weights
from the lowest levels of W_Λ. In all cases encountered so far
the highest weights which we may not find after projection of
W_Λ are those belonging to representations with one level only

(the trivial representation (00...0)) or a two-level representation. The latter ambiguity occurs only if the subalgebra contains A_1 and if the representation before projection has an even number of levels. Since the highest weights are above or at most on the middle level, it suffices to project only the weights of the restricted weight system W_Λ provided one can find the missing 1- or 2-dimensional representations by other means. Furthermore, a highest weight has all its coordinates (4) non-negative. Therefore, we do not need to keep track of any projected weight that has a negative component.

The projected weights with non-negative components then are assigned their levels using (6) as before. Their multiplicities are known because multiplicities of equal weights are added after the projection. Only the highest weights of irreducible representations of H are found on the highest level giving us part of our solution, namely, the "tallest" among the representations of H that we seek. Starting from each of these highest weights separately, we calculate its weight system and subtract its weights (with non-negative coordinates) from the projected system eliminating those weights which belong to these representations. In what remains of the projected system after subtraction, we look for what is now at the highest level and again conclude that these are also highest weights of irreducible representations of H, etc. Repeating this procedure we find the BR up to the one or two dimensional representations.

The missing representations are found by comparing the dimension $N(\Lambda)$ with $N(\Lambda')$ and the indices $I(\Lambda)$ with $I(\Lambda')$. Indeed, one must have

(10a) $N(\Lambda) = N(\Lambda') = N(\Lambda'_1) + N(\Lambda'_2) + \cdots$

and also [12]

(10b) $I(\Lambda) = qI(\Lambda') = q(I(\Lambda'_1) + I(\Lambda'_2) + \cdots),$

where the ratio $q = I(\Lambda)/I(\Lambda')$ is an invariant of the subalgebra independent of representations [12]. It is usually determined from the known BR for the lowest representation ω.

If the BR is at this stage complete, then (10a) and (10b) are satisfied. If (10a) does not hold but (10b) does, it implies that only some trivial representations whose I=0 are missing. However, if neither (10a) nor (10b) holds, the missing representations are (1) of A_1.

For convenience the number of missing representations is diminished by adding to the system W_Λ also the weights from one more level when T is odd.

3.   The Program

The computation was done on a CDC Cyber 74 computer at the Université de Montréal using the FORTRAN language. The program takes advantage of the 60-bit word size of this computer for economising on storage space so that in principle, calculations can be done for Lie algebras of rank up to 8 and representations of dimension up to about 10,000. More precisely, the number of weights of any restricted weight system generated must not exceed 5000. The high speed storage required is 121,000 words. The limiting factor is the time needed for calculating weight systems of high-dimensional representations. The highest dimension tested was in the calculation of the branching rules of the 6480-dimensional representation (1000010) of the Lie algebra E_7 reduced with respect to all its maximal semisimple subalgebras. This calculation took 1986 seconds of which 820 seconds was used for the calculation of the restricted weight system of the representation (1000010) of E_7. The results are presented in Table VII.

3.1. Program Description

The program BRULES is in two main parts: (i) the calculation of the restricted weight system W_Λ for a given irreducible represention of Z denoted by the highest weight Λ, and (ii) the calculation of the BR for the representation Λ into irreducible representations of any given semisimple subalgebra H by means of a projection matrix FIJ.

A brief description of the subprograms follows. The details of the method are described on COMMENT cards in the program. A complete listing of the simplest version of the program is given in the Appendix. This version shows how the calculation may be done for a single algebra-subalgebra pair (Z-H). In practice, since the weight system W_Λ for a given irreducible representation of Z need be calculated only once in order to find its BR into irreducible representations of any subalgebra H, the weight system W_Λ may be saved (e.g., on disc) after the calculation of part (i) is completed. Thereafter, only part (ii) needs to be run for the different subalgebras H which are of interest. This easy modification saves much time especially for the representations of high dimension. The results shown in Table VII are from the modified version of the program.

There is one restriction that the coordinates $(c_1, c_2, \ldots, c_n)$ of any weight must have values in the range $-50 \leq c_i \leq 50$ except for the case when $H = A_1$.

3.2. List of Subprograms

BRULES is the main program and controls the computation of the branching rules for a given irreducible representation HWZ of a (Z-H) pair. Calls: DATAH, DATAZ, PROJECT, WTSYSZ.

BLOCKDATA initializes the COMMON variables.

CALCWT generates the restricted weight system starting from a highest weight. Calls: ERR, PACK, UNPACK. Called from: PROJECT, WTSYSZ.

DATAH inputs the non-simple subalgebra H and the projection matrix FIJ. Called from: BRULES.

DATAZ inputs the algebra Z of rank ZRANK and the highest weight HWZ. Calls: ERR. Called from: BRULES.

ERR outputs error messages. Called from: CALCWT, DATAZ, PACK.

FREUD uses Freudenthal's algorithm for determining the multiplicities of weights in a restricted weight system and also calculates the dimension and the second order index of the representation. Calls: PACK, SPRD, UNPACK. Called from: PROJECT, WTSYSZ.

GETLL finds the lowest level of the restricted weight system and the number of gaps of the whole system of an irreducible representation of a simple Lie algebra. Called from: PROJECT, WTSYSZ.

GETLLH finds the lowest level of the restricted weight system and the number of gaps of the whole system of an irreducible representation of a non-simple Lie algebra. Calls: GETLL. Called from: PROJECT.

PACK packs a weight specified in an array into a single variable. Calls: ERR. Called from: CALCWT, FREUD, PROJECT.

PROJECT projects the restricted weight system (W,MU) of the irreducible representation of the simple algebra Z into a restricted weight system (PW, MUPW) of a reducible representation of its non-simple subalgebra H, and decomposes the projected system into irreducible components of H. Calls: CALCWT, FREUD, CETLL, CETLLH, PACK, SETUP, SORT1, SORT2, UNPACK. Called from:

BRULES.

SETUP sets up the Cartan matrix the scalar product matrix, the level vector and the adjoint representation of a simple Lie algebra. Called from: PROJECT, WTSYSZ.

SORT1 sorts the weight system (W,LV) into ascending order of levels. Called from: PROJECT, WTSYSZ.

SORT2 sorts the weight system (W,LV,MU) into ascending order of levels. Called from: PROJECT.

SPRD finds the scalar product of two vectors. Called from: FREUD.

UNPACK unpacks a weight specified in a single variable into an array. Called from: CALCWT, FREUD, PROJECT.

WTSYSZ calculates the restricted weight system (W,LV,MU) from a given highest weight HWZ of an irreducible representation of a simple Lie algebra Z. Calls: CALCWT, GETLL, FREUD, SETUP, SORT1. Called from: BRULES.

3.3 Description of the Data

The data for the program BRULES consists of:

a)   a parameter CHECK=1 if the dimension N and the index I of each irreducible representation as well as the ratio q are to be printed. Otherwise, CHECK=0.

b)   The algebra name Z, its rank ZRANK and the highest weight (HWZ(i), i=1,ZRANK).

c)   The subalgebra H which is specified in the form $H \equiv H_1 * H_2 * \ldots * H_{NSUB}$, where each H_i represents the name of a simple algebra followed by its rank.

d)   The projection matrix (FIJ(i,j),j=1, ZRANK,

i=1,HRANK), where HRANK is the rank of H.

The formats are as follows:

CARD	PARAMETERS	FORMAT	EXAMPLE
1.	CHECK	(I1)	1
2.	Z,ZRANK,HWZ	(A1,I1,1X,8I3)	C6 1 1 0 0 0 0
3.	H	(8(A1,I1,A1))	A1*C2
4.	FIJ	(8I3)	
5.			2 2 4 2 2 4
:	(HRANK cards)		1 2 1 2 1 2
:			0 0 1 1 2 1

3.4. Description of the Output

The output from the program BRULES is to the line printer.
The results show:

a)   the algebra Z of rank ZRANK and the given
highest weight HWZ.

b)   The subalgebra H and its irreducible components listed vertically.

c)   If a number and a dot (.) precede an
irreducible component of H, it indicates
the multiplicity of that component.

d)   As an option, the dimension N and the
index I of each irreducible representation as well as the ratio q may be
printed as an easy verification of (10a)
and (10b) that the BR is complete.

Example

The irreducible representation (110000) of the simple Lie
algebra C_6 which contains a reducible representation (4) (30) +

$(6)(11) + (2)(30) + (4)(11) + (2)(11) + (6)(10) + (0)(11) +$
$(4)(10) + (4)(10) + (2)(10) + (2)(10) + (0)(10)$ of the subalgebra
$A_1 * C_2$ is printed as

```
   C6                                  N           I              Q
      1  1  0  0  0  0                560          840

   ..............................................................

   A1*C2
         4  3  0                      100          610
         6  1  1                      112         1064
         2  3  0                       60          206
         4  1  1                       80          440
2  .     2  1  1                       96          272
         6  1  0                       28          238
         0  1  1                       16           24
2  .     4  1  0                       40          180
2  .     2  1  0                       24           44
         0  1  0                        4            2

   ..............................................................
                                      560         3080    .3667E+01
```

4.   Examples

In this section we describe examples of the construction of
the projection matrix and illustrate the computer procedure.
Two types of examples are considered. The first involves the
simple algebra of type A_2 and its two maximal simple subalgebras
both of which are of type A_1. The second type involves the algebra E_7 and some of its maximal subalgebras.

a.   The First Example: A_2.

There are two maximal semisimple subalgebras in A_2. Both
of them are simple of type A_1 and we consider them here simultaneously. The 3-dimensional representation $\omega \equiv (10)$ of A_2 contains then the A_1-representations

$$(11) \quad \begin{array}{ll} (10) \supset (2) & \text{(first case)} \\ (10) \supset (1) + (0) & \text{(second case).} \end{array}$$

We remark that in our notation the dimension of an A_1-represen-
tation (r) equals r + 1. In terms of the groups the first case
corresponds to $SU(3) \supset O(3)$ and the second one to $SU(3) \supset SU(2)$.

The projection matrix F is a 1×2 matrix because the subalgebra is of rank one and the algebra is of rank 2. In order
to find F one has to find the weights M_1 and M_2 ordered according
to a) and b) of Section 2. The problem is almost trivial. The
level structure of the ω_f representations in both cases are

$$(12) \quad \begin{array}{lll} \text{(1st case)} & \text{(2nd case)} & \\ M_1 = (2) \ldots\ldots\ldots\ldots & M_1 = (1) \ldots\ldots\ldots\ldots & \text{(top level)} \\ M_2 = (0) \ldots\ldots\ldots\ldots & M_2 = (0) \ldots\ldots\ldots\ldots & \text{(middle level)} \\ M_3 = (-2)\ldots\ldots\ldots\ldots & M_3 = (-1)\ldots\ldots\ldots\ldots & \text{(bottom level)} \end{array}$$

The weights M_1 and M_2 then determine the projection of the
vectors $v_1 = (10)$ and $v_2 = (01)$ and thus also the projection
of any weight. According to Table IV:

$$(13) \quad \begin{array}{ll} \text{(1st case)} & \text{(2nd case)} \\ (10) \ F = M_1 = (2) & (10) \ F = (1) \\ (01) \ F = M_1 + M_2 = (2) & (01) \ F = (1). \end{array}$$

Here in oder to preserve the matrix multiplication rules and
to avoid writing column-matrices, the row-matrices v_1 and v_2
are written in front of the projection matrix F_{ij}. Hence, a
general weight $(a_1 a_2)$ is projected as follows

$$(14) \qquad (a_1 a_2)\, F = (F_{11} a_1 + F_{12} a_2) = \begin{cases} (2a_1 + 2a_2) & \text{(1st case)} \\[2ex] (a_1 + a_2) & \text{(2nd case)} \end{cases}$$

Let us now calculate the BR for the 8-dimensional representation (11) of A_2. The number of levels of the weight system, $T+1$, equals 5. Hence, the restricted weight system W_Λ consists of the weights of the upper three levels. Using the weight computation algorithm one finds W_1 as

$$(15) \qquad \begin{array}{ll} (11) & \ldots \text{ level 1} \\ (-12)\ ,\ (2-1) & \ldots \text{ level 2} \\ (00)\ \ \ ,\ (00) & \ldots \text{ level 3} \end{array}$$

Application of F to W_Λ then produces the projected systems:

$$(16) \qquad \begin{array}{ll} \underline{\text{(1st case)}} & \underline{\text{(2nd case)}} \\ (11)F = (4) & (11)F = (2) \\ (2-1)F = (2) & (-12)F = (1) \\ (-12)F = (2) & (2-1)F = (1) \\ 2(00)F = 2(0) & 2(00)F = 2(0) \end{array}$$

Ordered according to levels, these weights can be written as

$$(17) \qquad \begin{array}{ll} (4) & (2) \\ (2)\ ,\ (2) & (1)\ ,\ (1) \\ (0)\ ,\ (0) & (0)\ ,\ (0) \end{array}$$

respectively. On the top level we find the highest weight of an irreducible representation. These are (4) and (2) respectively if the restricted weights systems corresponding to each of them are subtracted from (17) one is left with

$$(18) \qquad \begin{array}{ll} (2) & (1)\ ,\ (1) \\ (0) & (0) \end{array}$$

instead of (17). Clearly in the first case the two weights (2)
and (0) form W_Λ of the representation (2). Thus in this case
the BR reads

(19)   (10) ⊃ (4) + (2).

In the second case, the top level in (18) contains two weights
(1) and (1). Hence, both of them are highest weights of two
dimensional representations. When these are subtracted from (18)
the only remaining weight is (0), the highest weight of 1-dim-
ensional trivial representation. Finally, the BR of the second
case is

(20)   (10) ⊃ (2) + (1) + (1) + (0).

b.   The Second Example: E_7

The exceptional simple Lie algebra E_7 has 12 maximal semisimple algebras. They are types A_7, E_6, A_2, A_1*F_4, G_2*C_3,
A_1*G_2, A_1*A_1, A_1*D_6, A_2*A_5, $A_1*A_3*A_3$, and two different subalgebras both of type A_1. In view of the recent applications
[6] an interesting subalgebra is A_2*A_5. Let us demonstrate
here the way the projection matrix F is constructed in this
case.

First, one has to know the BR for the lowest nontrivial
representation (0000010) of E_7 to the subalgebra A_2*A_5. It
reads [11] in our conventions

(21)   (0000010) ⊃ (00)(00100) + (10)(10000) + (01)(00001)

Next we need 7 weights M_1, M_2,...,M_7 of the reducible representation of A_2*A_5 on the right side of (21) ordered according to
the levels and some lexicographical ordering on each level. For
that we need first to find out how "tall" are the three irreducible weight systems in (21). Using (6) we find the following
level structure of the reducible weight system.

(22)

1st level	(00)(00100)
2nd level	(00)(01-110) , (10)(10000) , (01)(00001)
3rd level	(00)(1-1010) , (00)(010-11) , (-11)(10000),...

Putting

(23)

$$
\begin{aligned}
M_1 &= (00)(00100) \\
M_2 &= (10)(10000) \\
M_3 &= (00)(01-110) \\
M_4 &= (01)(00001) \\
M_5 &= (-11)(10000) \\
M_6 &= (00)(1-1010) \\
M_7 &= (00)(01-101),
\end{aligned}
$$

we obtain the action of F on the vectors $v_1 = (1000000)$, $v_2 = (0100000),\ldots, v_7 = (0000001)$ using Table VI as

(24)

$$
\begin{aligned}
v_1F &= (00)(10001) \\
v_2F &= (01)(11001) \\
v_3F &= (11)(11011) \\
v_4F &= (10)(11010) \\
v_5F &= (10)(10100) \\
v_6F &= (00)(00100) \\
v_7F &= (01)(10010).
\end{aligned}
$$

Hence, a general weight $(a_1a_2\ldots a_7)$ of E_7 is projected as follows

$$
\begin{aligned}
(a_1a_2 \ \cdots \ a_7) \ F &= (F_{1i}a_i, \ F_{2i}a_i)(F_{3i}a_i, \ F_{4i}a_i,\ldots,F_{7i}a_i) \\
\text{(25)} \quad &= (a_3 + a_4 + a_5, \ a_2 + a_3 + a_7)(a_1 + a_2 + a_3 + a_4 + a_5 + a_7, \\
&\quad a_2 + a_3 + a_4, \ a_5 + a_6, \ a_3 + a_4 + a_7, \ a_1 + a_2 + a_3).
\end{aligned}
$$

In order to illustrate the BR computation, let us consider the first few steps of the BR calculation for the irreducible representation (1000010) of E_7 (dimensions 6480) reduced with respect to A_2*A_5. The top level of W_Λ consists of the weight

V_1 = (1000010), the second level contains V_2 = (-1100010) and
V_3 = (10001-10), to the third level belong V_4 = (0-110010),
V_5 = (-11001-10), V_6 = (1001-100); the multiplicities of all
these weights equal 1.   Then

$$V_1F = (00)(10101)$$
$$V_2F = (01)(01100)$$
$$V_3F = (10)(20001)$$
(26) $$V_4F = (10)(00110)$$
$$V_5F = (11)(11000)$$
$$V_6F = (00)(11-111).$$

The highest weight V_1 is projected always into a highest weight,
i.e., into (00)(10101). Among the weights V_2F, V_3F,...,V_6F only
the last one belongs to the weight system W_Λ of (00)(10101) so
that after the weights of these systems are subtracted from (26),
we are left with weights V_2F, V_3F, V_4F, and V_5F. It is a matter
of an elementary inspection of the weight algorithm in order to
conclude that none of these four weights belongs to the weight
system of any other. Hence, each of them must be a highest
weight of one of the irreducible representations $A_2{}^*A_5$. Consequently we have established that

(27) $(1000010) \supset (00)(10101) + (01)(01100) + (10)(20001) +$
         $+ (10)(00110) + (11)(11000) + \ldots$

Furthermore, from these first 6 irreducible components it
follows immediately that the BR (27) must contain as well some
further irreducible representations. Indeed, any E_7 representation is self-contragredient [3], hence, it contains either
representations which are self-contragredient or pairs of
mutually contragredient representations of $A_2{}^*A_5$. Among the
irreducible representations in (27) only (00)(10101) and the
pair (01)(01100) + (10)(00110) are self-contragredient. Thus,
finally we have found by an easy hand computation a considerable
part of the BR for the 6480-dimensional representation (10000010)

of E_7 with respect to the subalgebra A_2*A_5:

(28)   (1000010) ⊃ (00)(10101) + (01)(01100) + (10)(00110) +

        + (10)(20001) + (01)(10002) + (11)(11000) + (11)(00011) + ...

The remaining part of this BR is found in Table VII.

References

[1] V. K. Agrawala and J. G. Belinfante, *Weight diagrams for Lie group representations: A computer implementation of Freudenthal's algorithm in Algol and Fortran*, BIT 9 (1969), 301-314.

[2] V. I. Arnold, *Critical points of smooth functions and their normal forms*, Uspekchi Mat. Nauk 30 (1975), 3

[3] A. K. Bose and J. Patera, *Classification of finite-dimensional irreducible representations of connected complex semisimple Lie groups*, J. Math. Phys. 11 (1970), 2231-2234.

[4] R. W. Carter, *Simple Groups of Lie Type*, Wiley, New York, 1972.

[5] E. B. Dynkin, *Semisimple subalgebras of semisimple Lie algebras*, Mat. Sbornik N.S. 30 (1952), 349-462; translated in Amer. Math. Soc. Transl. Ser. 2, 6 (1957), 111-244.

[6] F. Gursey and P. Sikivie, *E_7 as a universal gauge group*, Phys. Rev. Letts. 36 (1976), 775-778.

[7] R. C. King, *Branching rules for classical Lie groups using tensor and spinor methods*, J. Phys. A 8 (1975), 429-449.

[8] E. Loebl, (editor), *Group Theory and its Applications*, Vol. I, II and III, Academic Press, New York 1968, 1971 and 1975.

[9] W. McKay, J. Patera, and R. T. Sharp, *Branching rules and Clebsch-Gordan coefficients for E_8*, J. Math. Phys. 17 (1976), 1371-1375.

[10] A. Navon, and J. Patera, *Embedding of simple Lie group into a simple Lie group and branching rules*, J. Math. Phys. 8, (1967), 489-493.

[11] J. Patera and D. Sankoff, *Tables of Branching Rules for Representations of Simple Lie Algebras*, Presses Université de Montréal, Montreal, 1973.

[12] J. Patera, R.T. Sharp and P. Winternitz, *Higher indices of group representations*, J. Math. Phys. 17 (1976), to appear.

Centre de recherches mathématiques
Université de Montréal
Montréal, Québec, Canada

Table 1

Numbering of Simple Roots.

Table II

The Level Vectors $R = (r_1, r_2, \ldots, r_n)$ of

Representations of Simple Lie Algebras

A_n: $\quad R = (n \cdot 1, (n-1) \cdot 2, \ldots, 2 \cdot (n-1), 1 \cdot n)$

B_n: $\quad R = (1 \cdot 2n, 2 \cdot (2n-1), \ldots, (n-k+1)(n+k), \ldots, (n-2)(n+3), (n-1)$
$\quad\quad\quad (n+2), 1/2n(n+1))$

C_n: $\quad R = (1 \cdot (2n-1), 2(2n-2), \ldots, (n-k+1)(n+k-1), \ldots, (n-2), (n+2),$
$\quad\quad\quad (n-1)(n+1), n^2)$

D_n: $\quad R = (1 \cdot (2n-2), 2(2n-3), \ldots, (n-k+1)(n+k-2), \ldots, (n-3)(n+2),$
$\quad\quad\quad (n-2)(n+1), \; 1/2n(n-1))$

E_6: $\quad R = (16, 30, 42, 30, 16, 22)$

E_7: $\quad R = (34, 66, 96, 75, 52, 27, 49)$

E_8: $\quad R = (58, 114, 168, 220, 270, 186, 92, 136)$

F_4: $\quad R = (22, 42, 30, 16)$

G_2: $\quad R = (10, 6)$

Table III

Scalar Product Matrices (g_{ij})

$$A_n \quad \frac{1}{n+1} \begin{bmatrix}
1.n & 1.(n-1) & 1.(n-2) & \cdots & 1.2 & 1.1 \\
1.(n-1) & 2.(n-1) & 2.(n-2) & \cdots & 2.2 & 2.1 \\
1.(n-2) & 2.(n-2) & 3.(n-2) & \cdots & 3.2 & 3.1 \\
& & & & & \\
1.2 & 2.2 & 3.2 & \cdots & (n-1).2 & (n-1).1 \\
1.1 & 2.1 & 3.1 & \cdots & (n-1).1 & n.1
\end{bmatrix}$$

$$B_n \quad \frac{1}{2} \begin{bmatrix}
2 & 2 & 2 & \cdots & 2 & 1 \\
2 & 4 & 4 & \cdots & 4 & 2 \\
2 & 4 & 6 & \cdots & 6 & 3 \\
\cdots\cdots\cdots\cdots & & & \cdots & \cdots\cdots\cdots\cdots & \\
2 & 4 & 6 & \cdots & 2(n-1) & n-1 \\
1 & 2 & 3 & \cdots & n-1 & \frac{n}{2}
\end{bmatrix}$$

$$C_n \quad \frac{1}{4} \begin{bmatrix}
1 & 1 & 1 & \cdots & 1 & 1 \\
1 & 2 & 2 & \cdots & 2 & 2 \\
1 & 2 & 3 & \cdots & 3 & 3 \\
\cdots\cdots\cdots\cdots & & & \cdots & \cdots\cdots\cdots\cdots & \\
1 & 2 & 3 & \cdots & n-1 & n-1 \\
1 & 2 & 3 & \cdots & n-1 & n
\end{bmatrix}$$

$$D_n \quad \frac{1}{2} \begin{bmatrix}
2 & 2 & 2 & \cdots & 2 & 1 & 1 \\
2 & 4 & 4 & \cdots & 4 & 2 & 2 \\
2 & 4 & 6 & \cdots & 6 & 3 & 3 \\
\cdots\cdots\cdots\cdots & & & \cdots & \cdots\cdots\cdots\cdots & & \\
2 & 4 & 6 & \cdots & 2(n-2) & n-2 & n-2 \\
1 & 2 & 3 & \cdots & n-2 & \frac{n}{2} & \frac{n-2}{2} \\
1 & 2 & 3 & \cdots & n-2 & \frac{n-2}{2} & \frac{n}{2}
\end{bmatrix}$$

$$E_6 \quad \frac{1}{3} \begin{bmatrix}
4 & 5 & 6 & 4 & 2 & 3 \\
5 & 10 & 12 & 8 & 4 & 6 \\
6 & 12 & 18 & 12 & 6 & 9 \\
4 & 8 & 12 & 10 & 5 & 6 \\
2 & 4 & 6 & 5 & 4 & 3 \\
3 & 6 & 9 & 6 & 3 & 6
\end{bmatrix}$$

(continued on following page)

Table III   (continued)

$$E_7 \quad \frac{1}{2} \begin{bmatrix} 4 & 6 & 8 & 6 & 4 & 2 & 4 \\ 6 & 12 & 16 & 12 & 8 & 4 & 8 \\ 8 & 16 & 24 & 18 & 12 & 6 & 12 \\ 6 & 12 & 18 & 15 & 10 & 5 & 9 \\ 4 & 8 & 12 & 10 & 8 & 4 & 6 \\ 2 & 4 & 6 & 5 & 4 & 3 & 3 \\ 4 & 8 & 12 & 9 & 6 & 3 & 7 \end{bmatrix}$$

$$E_8 \quad \begin{bmatrix} 2 & 3 & 4 & 5 & 6 & 4 & 2 & 3 \\ 3 & 6 & 8 & 10 & 12 & 8 & 4 & 6 \\ 4 & 8 & 12 & 15 & 18 & 12 & 6 & 9 \\ 5 & 10 & 15 & 20 & 24 & 16 & 8 & 12 \\ 6 & 12 & 18 & 24 & 30 & 20 & 10 & 15 \\ 4 & 8 & 12 & 16 & 20 & 14 & 7 & 10 \\ 2 & 4 & 6 & 8 & 10 & 7 & 4 & 5 \\ 3 & 6 & 9 & 12 & 15 & 10 & 5 & 8 \end{bmatrix}$$

$$G_2 \quad \frac{1}{3} \begin{bmatrix} 6 & 3 \\ 3 & 2 \end{bmatrix}$$

$$F_4 \quad \begin{bmatrix} 2 & 3 & 2 & 1 \\ 3 & 6 & 4 & 2 \\ 2 & 4 & 3 & \frac{3}{2} \\ 1 & 2 & \frac{3}{2} & 1 \end{bmatrix}$$

```
C2 - A1
3  4

C2 - A1*A1
1  1
0  1

G2 - A2
1  1
1  0

G2 - A1
10  6

G2 - A1*A1
1  1
3  1

A3 - A2
1  1  0
0  0  1

A3 - C2
1  0  1
0  1  0

A3 - A1*A1
1  0  1
1  2  1

B3 - G2
0  1  0
1  0  1

B3 - A3
0  1  1
1  0  0
0  1  0

B3 - A1*A1*A1
1  1  0
1  1  1
0  2  1

C3 - A2
1  1  2
0  1  0

C3 - A1
5  8  9

C3 - A1*A1
1  0  1
2  4  4

C3 - A1*C2
0  0  1
1  0  0
0  1  1

A4 - A3
1  0  0  0
0  1  1  0
0  0  0  1

A4 - C2
0  2  2  0
1  0  0  1

A4 - A1*A2
0  1  1  0
1  1  0  0
0  0  1  1

B4 - D4
1  0  0  0
0  1  0  0
0  0  1  0
0  0  1  1

B4 - A1
8  14  18  10

B4 - A1*A1
2  2  4  1
2  4  4  3

B4 - A1*A1*C2
0  1  1  0
0  1  1  1
0  0  2  1
1  1  0  0

B4 - A1*A3
0  0  2  1
0  1  1  0
1  0  0  0
0  1  1  1


C4 - A3
1  1  0  0
0  0  1  2
0  1  1  0

C4 - A1
7  12  15  16

C4 - A1*C3
0  0  0  1
1  0  0  0
0  1  0  0
0  0  1  1

C4 - C2*C2
0  1  1  0
0  0  0  1
1  1  0  0
0  0  1  1

C4 - A1*A1*A1
1  0  1  2
1  2  1  2
1  2  3  2

D4 - A3
1  1  1  0
0  0  0  1
0  1  0  0

D4 - B3
0  0  0  1
0  1  0  0
1  0  1  0

D4 - A2
1  0  1  1
1  3  1  1

D4 - A1*C2
1  2  0  1
1  0  0  1
0  1  1  0

D4 - A1*A1*A1*A1
1  1  0  0
1  1  1  1
0  1  1  0
0  1  0  1

F4 - B4
0  1  1  0
1  0  0  0
0  1  0  0
0  0  1  1

F4 - A1
22  42  30  16

F4 - A1*G2
4  4  4  2
0  1  1  0
1  1  0  1

F4 - A1*A3
2  2  2  1
0  1  1  1
1  1  0  0
0  1  1  0

F4 - A1*C3
2  3  2  1
0  0  0  1
0  0  1  0
0  1  0  0

F4 - A2*A2
0  0  1  1
0  2  1  0
1  2  1  1
1  1  1  0

A5 - A4
1  0  0  0  0
0  1  0  0  0
0  0  1  1  0
0  0  0  0  1

A5 - A3
0  1  0  1  0
1  0  0  0  1
0  1  2  1  0

A5 - C3
1  0  0  0  1
0  1  0  1  0
0  0  1  0  0

A5 - A2
0  1  3  2  2
2  2  0  1  0


A5 - A1*A2
1  0  1  0  1
1  2  1  0  0
0  0  1  2  1

A5 - A1*A3
0  0  1  0  0
1  0  0  0  0
0  1  1  1  0
0  0  0  0  1

A5 - A2*A2
0  1  0  0  0
0  0  1  1  0
1  1  1  0  0
0  0  0  1  1

B5 - D5
1  0  0  0  0
0  1  0  0  0
0  0  1  0  0
0  0  0  1  0
0  0  0  1  1

B5 - A1
10  18  24  28  15

B5 - A1*A1*B3
0  0  1  1  0
0  0  1  1  1
1  0  0  0  0
0  1  1  0  0
0  0  0  2  1

B5 - A1*D4
0  0  0  2  1
1  0  0  0  0
0  1  0  0  0
0  0  1  1  0
0  0  1  1  1

B5 - A3*C2
0  0  1  1  0
1  1  0  0  0
0  0  1  1  1
0  0  0  2  1
0  1  1  0  0

C5 - A4
1  1  0  0  0
0  0  1  1  0
0  0  0  1  2
0  1  1  0  0

C5 - A1
9  16  21  24  25

C5 - A1*C4
0  0  0  0  1
1  0  0  0  0
0  1  0  0  0
0  0  1  0  0
0  0  0  1  1

C5 - C2*C3
0  0  1  1  0
0  0  0  0  1
1  0  0  0  0
0  1  1  0  0
0  0  0  1  1

C5 - A1*C2
1  0  1  0  1
0  0  2  4  4
1  2  1  0  0

D5 - A4
1  1  0  0  0
0  0  1  0  1
0  0  0  1  0
0  1  1  0  0

D5 - B4
1  0  0  0  0
0  1  0  0  0
0  0  1  0  0
0  0  0  1  1

D5 - C2
2  2  4  1  1
0  1  0  1  1

D5 - A1*A1*A3
0  1  1  0  0
0  1  1  1  1
0  0  1  1  0
1  1  0  0  0
0  0  1  0  1

D5 - A1*B3
0  0  0  1  1
1  0  0  0  0
0  1  0  0  0
0  0  2  1  1


D5 - C2*C2
0  0  2  1  1
1  1  0  0  0
0  0  0  1  1
0  1  1  0  0

A6 - A5
1  0  0  0  0  0
0  1  0  0  0  0
0  0  1  1  0  0
0  0  0  0  1  0
0  0  0  0  0  1

A6 - B3
1  0  0  0  0  1
0  1  0  0  1  0
0  0  2  2  0  0

A6 - A1*A4
0  0  1  1  0  0
1  0  0  0  0  0
0  1  1  0  0  0
0  0  0  1  1  0
0  0  0  0  0  1

A6 - A2*A3
0  1  1  0  0  0
0  0  0  1  1  0
1  1  0  0  0  0
0  0  1  1  0  0
0  0  0  0  1  1

B6 - D6
1  0  0  0  0  0
0  1  0  0  0  0
0  0  1  0  0  0
0  0  0  1  0  0
0  0  0  0  1  1

B6 - A1
12  22  30  36  40  21

B6 - A1*A1*B4
0  0  0  1  1  1
0  0  1  1  1  1
1  0  0  0  0  0
0  1  0  0  0  0
0  0  1  1  0  0
0  0  0  0  2  1

B6 - A1*D5
0  0  0  0  2  1
1  0  0  0  0  0
0  1  0  0  0  0
0  0  1  0  0  0
0  0  0  1  1  0
0  0  0  1  1  1

B6 - A3*B3
0  0  0  1  1  0
0  1  1  0  0  0
0  0  0  1  1  1
1  1  0  0  0  0
0  0  1  1  0  0
0  0  0  0  2  1

B6 - C2*D4
0  0  0  0  2  1
0  0  1  1  0  0
1  0  0  0  0  0
0  1  1  0  0  0
0  0  0  1  1  0
0  0  0  1  1  1

C6 - A5
1  1  0  0  0  0
0  0  1  1  0  0
0  0  0  0  1  2
0  0  0  1  0  0
0  1  1  0  0  0

C6 - A1
11  20  27  32  35  36

C6 - A1*A3
1  0  1  0  1  2
0  0  1  2  1  2
1  2  1  0  0  0
0  0  1  2  3  2

C6 - A1*C2
2  2  4  2  2  4
1  2  1  2  1  2
0  0  1  1  2  1


C6 - A1*C5
0  0  0  0  0  1
1  0  0  0  0  0
0  1  0  0  0  0
0  0  1  0  0  0
0  0  0  1  0  0
0  0  0  0  1  1

C6 - C2*C4
0  0  0  1  1  0
0  0  0  0  0  1
1  0  0  0  0  0
0  1  0  0  0  0
0  0  1  1  0  0
0  0  0  0  1  1

C6 - C3*C3
0  1  1  0  0  0
0  0  0  1  1  0
0  0  0  0  0  1
1  1  0  0  0  0
0  0  1  1  0  0
0  0  0  0  1  1

D6 - A5
1  1  0  0  0  0
0  0  1  1  0  0
0  0  0  0  0  1
0  0  0  0  1  0
0  1  1  0  0  0

D6 - B5
1  0  0  0  0  0
0  1  0  0  0  0
0  0  1  0  0  0
0  0  0  1  0  0
0  0  0  0  1  1

D6 - A1*C3
1  2  1  2  0  1
1  0  1  1  0  0
0  1  1  0  0  1
0  1  1  0  0  1

D6 - A1*A1*D4
0  0  1  1  1  1
0  0  1  1  1  1
1  0  0  0  0  0
0  1  1  0  0  0
0  0  0  1  0  1
0  0  0  1  0  1

D6 - A3*A3
0  0  1  0  0  0
0  1  1  0  0  1
0  0  1  1  1  1
0  0  1  1  1  0
1  1  0  0  0  0
0  0  1  1  0  1

D6 - A1*B4
0  0  0  0  1  1
1  0  0  0  0  0
0  1  0  0  0  0
0  0  1  0  0  0
0  0  0  2  1  1

D6 - C2*B3
0  0  0  1  1  0
0  0  1  1  0  0
1  0  0  0  0  0
0  1  1  0  0  0
0  0  0  2  1  1

D6 - A1*A1*A1
2  4  6  6  4  4
1  2  1  2  1  0
1  0  1  2  1  0

E6 - D5
0  1  1  0  0
0  0  0  0  0
0  0  1  0  0
0  0  0  1  1
1  1  0  0  0

E6 - C4
0  1  0  1  0  0
1  0  0  0  1  0
0  1  2  1  0  0
0  0  0  0  0  1

E6 - F4
0  0  0  0  0  1
0  0  1  0  0  0
0  1  0  1  0  0
1  0  0  0  1  0
```

```
E6  -  A2
2  2  5  5  2  1
2  5  5  2  2  4

E6  -  G2
0  1  0  1  0  1
2  2  5  2  2  1

E6  -  A2*G2
1  2  1  0  0  1
0  0  1  2  1  1
0  1  1  1  0  0
1  0  1  0  1  1

E6  -  A1*A5
0  1  1  1  0  1
0  1  1  0  0  0
0  0  0  1  1  0
0  0  1  0  0  1
1  1  0  0  0  0
0  0  1  1  0  0

E6  -  A2*A2*A2
1  1  1  1  1  1
0  1  1  1  0  0
0  0  1  1  1  0
0  1  1  0  0  1
0  0  1  1  0  1
1  1  1  0  0  0

A7  -  A6
1  0  0  0  0  0  0
0  1  0  0  0  0  0
0  0  1  0  0  0  0
0  0  0  1  1  0  0
0  0  0  0  0  1  0
0  0  0  0  0  0  1

A7  -  C4
1  0  0  0  0  0  1
0  1  0  0  0  1  0
0  0  1  0  1  0  0
0  0  0  1  0  0  0

A7  -  D4
1  0  0  0  0  0  1
0  1  0  0  0  1  0
0  0  1  0  1  0  0
0  0  1  2  1  0  0

A7  -  A1*A3
1  0  1  0  1  0  1
1  2  1  0  0  0  0
0  0  1  2  1  0  0
0  0  0  0  1  2  1

A7  -  A1*A5
0  0  0  1  0  0  0
1  0  0  0  0  0  0
0  1  0  0  0  0  0
0  0  0  1  1  0  0
0  0  0  0  0  1  0
0  0  0  0  0  0  1

A7  -  A2*A4
0  0  1  1  0  0  0
0  0  0  0  1  0  0
1  0  0  0  0  0  0
0  1  1  0  0  0  0
0  0  0  1  1  1  0
0  0  0  0  0  0  1

A7  -  A3*A3
0  1  1  0  0  0  0
0  0  0  1  0  0  0
0  0  0  0  1  1  0
1  1  0  0  0  0  0
0  0  1  1  1  0  0
0  0  0  0  0  1  1

B7  -  D7
1  0  0  0  0  0  0
0  1  0  0  0  0  0
0  0  1  0  0  0  0
0  0  0  1  0  0  0
0  0  0  0  1  0  0
0  0  0  0  0  1  0
0  0  0  0  0  1  1

B7  -  A3
1  0  1  1  0  2  1
0  1  2  1  3  2  1
1  2  1  3  2  2  1

B7  -  A1
14  26  36  44  50  54  28

B7  -  A1*C2
2  2  4  2  2  4  1
0  0  2  2  4  4  3
1  2  1  2  1  1  0


B7  -  A1*D6
0  0  0  0  0  2  1
1  0  0  0  0  0  0
0  1  0  0  0  0  0
0  0  1  0  0  0  0
0  0  0  1  0  0  0
0  0  0  0  1  0  0
0  0  0  0  1  1  0
0  0  0  0  1  1  1

B7  -  A1*A1*B5
0  0  0  0  1  1  0
0  0  0  0  1  1  1
1  0  0  0  0  0  0
0  1  0  0  0  0  0
0  0  1  0  0  0  0
0  0  0  1  1  0  0
0  0  0  0  0  2  1

B7  -  C2*D5
0  0  0  0  0  2  1
0  0  0  1  1  0  0
1  0  0  0  0  0  0
0  1  0  0  0  0  0
0  0  1  1  0  0  0
0  0  0  0  1  0  0
0  0  0  0  1  1  1

B7  -  A3*B4
0  0  0  0  1  1  0
0  0  1  1  0  0  0
0  0  0  0  1  1  1
1  0  0  0  0  0  0
0  1  1  0  0  0  0
0  0  0  1  1  0  0
0  0  0  0  0  2  1

B7  -  B3*D4
0  1  1  0  0  0  0
0  0  0  1  1  0  0
0  0  0  0  0  2  1
1  1  0  0  0  0  0
0  0  1  1  0  0  0
0  0  0  0  1  1  0
0  0  0  0  1  1  1

C7  -  A6
1  1  0  0  0  0  0
0  0  1  1  0  0  0
0  0  0  0  1  1  0
0  0  0  0  0  1  2
0  0  0  1  1  0  0
0  1  1  0  0  0  0

C7  -  A1
13  24  33  40  45  48  49

C7  -  A1*B3
1  0  1  0  1  0  1
1  2  1  0  1  0  0
0  0  1  2  1  0  0
0  0  0  0  2  4  4

C7  -  A1*C6
0  0  0  0  0  0  1
1  0  0  0  0  0  0
0  1  0  0  0  0  0
0  0  1  0  0  0  0
0  0  0  1  0  0  0
0  0  0  0  1  0  0
0  0  0  0  0  1  1

C7  -  C2*C5
0  0  0  0  1  1  0
0  0  0  0  0  0  1
1  0  0  0  0  0  0
0  1  0  0  0  0  0
0  0  1  0  0  0  0
0  0  0  1  1  0  0
0  0  0  0  0  1  1

C7  -  C3*C4
0  0  1  1  0  0  0
0  0  0  0  1  1  0
0  0  0  0  0  0  1
1  0  0  0  0  0  0
0  1  1  0  0  0  0
0  0  0  1  1  0  0
0  0  0  0  0  1  1

D7  -  A6
1  1  0  0  0  0  0
0  0  1  1  0  0  0
0  0  0  0  1  1  0
0  0  0  0  0  0  1
0  0  0  1  1  0  0
0  1  1  0  0  0  0


D7  -  B6
1  0  0  0  0  0  0
0  1  0  0  0  0  0
0  0  1  0  0  0  0
0  0  0  1  0  0  0
0  0  0  0  1  0  0
0  0  0  0  0  1  1

D7  -  C3
0  1  0  1  0  1  1
1  0  0  1  3  1  1
0  1  2  1  0  0  0

D7  -  C2
0  2  2  6  4  3  3
2  2  3  1  3  1  1

D7  -  G2
1  0  0  1  0  1  1
0  3  4  3  5  1  1

D7  -  A1*A1*D5
0  0  0  1  1  0  0
0  0  0  1  1  1  1
1  0  0  0  0  0  0
0  1  0  0  0  0  0
0  0  1  1  0  0  0
0  0  0  0  1  1  0
0  0  0  0  1  0  1

D7  -  A3*D4
0  0  0  1  1  0  0
0  1  1  0  0  0  0
0  0  0  1  1  1  1
1  1  0  0  0  0  0
0  0  1  1  0  0  0
0  0  0  0  1  1  0
0  0  0  0  1  0  1

D7  -  A1*B5
0  0  0  0  0  1  1
1  0  0  0  0  0  0
0  1  0  0  0  0  0
0  0  1  0  0  0  0
0  0  0  1  0  0  0
0  0  0  0  2  1  1

D7  -  C2*B4
0  0  0  0  0  1  1
0  0  0  1  1  0  0
1  0  0  0  0  0  0
0  1  0  0  0  0  0
0  0  1  1  0  0  0
0  0  0  0  2  1  1

D7  -  B3*B3
1  1  0  0  0  0  0
0  0  1  1  0  0  0
0  0  0  0  2  1  1
0  1  1  0  0  0  0
0  0  0  1  1  0  0
0  0  0  0  0  1  1

E7  -  A7
0  1  1  1  0  0  1
0  0  0  0  1  1  0
0  0  1  1  0  0  0
1  1  0  0  0  0  1
0  0  1  0  0  0  1
0  0  0  1  1  0  0
0  1  1  0  0  0  0

E7  -  E6
0  0  0  0  1  1  0
0  0  1  1  0  0  0
0  1  0  0  0  0  0
0  0  1  0  0  0  1
0  0  0  1  1  0  0
1  0  0  0  0  0  0

E7  -  A2
4  9  11  10  6  6  7
4  6  11  7  6  0  4

E7  -  A1
34  66  96  75  52  27  49

E7  -  A1
26  50  72  57  40  21  37

E7  -  A1*F4
0  0  2  1  2  1  1
1  0  0  0  0  0  0
0  1  1  0  0  0  0
0  0  0  1  1  0  1
0  0  1  1  0  1  0

E7  -  G2*C3
0  1  0  1  1  1  0
1  2  1  1  2  1  0
0  1  1  1  0  1  0
1  0  1  1  0  0  0
0  1  1  0  0  0  1


E7  -  A1*G2
2  4  4  5  4  1  3
0  1  1  0  1  1  1
2  2  4  4  1  0  1

E7  -  A1*A1
4  10  18  12  8  6  8
6  10  12  11  8  3  7

E7  -  A1*D6
0  1  1  1  0  0  1
0  0  1  1  0  0  0
1  1  0  0  0  0  0
0  0  1  0  0  0  1
0  0  0  1  1  0  0
0  1  1  0  0  0  0
0  0  0  0  0  1  0

E7  -  A2*A5
0  0  1  1  1  0  0
0  1  1  0  0  0  1
1  1  1  1  1  0  1
0  1  1  1  0  0  0
0  0  0  0  1  1  0
0  0  1  1  0  0  1
1  1  1  0  0  0  1

E7  -  A1*A3*A3
0  1  2  1  0  0  1
0  1  1  1  1  1  1
1  1  1  1  0  0  0
0  0  1  1  1  1  0
1  1  1  0  0  1  0
0  1  1  1  1  0  0

A8  -  A7
1  0  0  0  0  0  0  0
0  1  0  0  0  0  0  0
0  0  1  0  0  0  0  0
0  0  0  1  1  0  0  0
0  0  0  0  0  1  0  0
0  0  0  0  0  0  1  0
0  0  0  0  0  0  0  1

A8  -  B4
1  0  0  0  0  0  0  1
0  1  0  0  0  0  1  0
0  0  1  0  0  0  1  0
0  0  0  2  2  0  0  0

A8  -  A2*A2
1  0  1  0  1  1  1  0
0  1  1  0  1  1  0  1
1  2  1  2  1  1  0  1
0  0  1  1  2  1  2  1

A8  -  A1*A6
0  0  0  1  1  0  0  0
1  0  0  0  0  0  0  0
0  1  0  0  0  0  0  0
0  0  1  1  0  0  0  0
0  0  0  0  1  1  0  0
0  0  0  0  0  0  1  0
0  0  0  0  0  0  0  1

A8  -  A2*A5
0  0  0  1  1  0  0  0
0  0  1  0  0  1  0  0
1  0  0  0  0  0  0  0
0  1  1  0  0  0  0  0
0  0  0  1  1  0  0  0
0  0  0  0  1  1  0  0
0  0  0  0  0  0  1  1

A8  -  A3*A4
0  1  1  0  0  0  0  0
0  0  0  1  1  0  0  0
0  0  0  0  0  1  1  0
1  1  0  0  0  0  0  0
0  0  1  1  0  0  0  0
0  0  0  0  1  1  0  0
0  0  0  0  0  1  1  1

B8  -  D8
1  0  0  0  0  0  0  0
0  1  0  0  0  0  0  0
0  0  1  0  0  0  0  0
0  0  0  1  0  0  0  0
0  0  0  0  1  0  0  0
0  0  0  0  0  1  0  0
0  0  0  0  0  0  1  0
0  0  0  0  0  0  0  1

B8  -  A1
16  30  42  52  60  66  70  36
```

```
B8 - A1*D7
0  0  0  0  0  0  2  1
1  0  0  0  0  0  0  0
0  1  0  0  0  0  0  0
0  0  1  0  0  0  0  0
0  0  0  1  0  0  0  0
0  0  0  0  1  0  0  0
0  0  0  0  0  1  1  0
0  0  0  0  0  1  1  1

B8 - A1*A1*B6
0  0  0  0  0  1  1  0
0  0  0  0  0  1  1  1
1  0  0  0  0  0  0  0
0  1  0  0  0  0  0  0
0  0  1  0  0  0  0  0
0  0  0  1  0  0  0  0
0  0  0  0  1  1  0  0
0  0  0  0  0  2  1  1

B8 - C2*D6
0  0  0  0  0  0  2  1
0  0  0  0  1  1  0  0
1  0  0  0  0  0  0  0
0  1  0  0  0  0  0  0
0  0  1  0  0  0  0  0
0  0  0  1  1  0  0  0
0  0  0  0  0  1  1  0
0  0  0  0  0  1  1  1

B8 - A3*B5
0  0  0  0  0  1  1  0
0  0  0  1  1  0  0  0
0  0  0  0  0  1  1  1
1  0  0  0  0  0  0  0
0  1  0  0  0  0  0  0
0  0  1  1  0  0  0  0
0  0  0  0  1  1  0  0
0  0  0  0  0  0  2  1

B8 - B3*D5
0  0  1  1  0  0  0  0
0  0  0  0  1  1  0  0
0  0  0  0  0  0  2  1
1  0  0  0  0  0  0  0
0  1  1  0  0  0  0  0
0  0  0  1  1  0  0  0
0  0  0  0  0  1  1  0
0  0  0  0  0  1  1  1

B8 - B4*D4
1  1  0  0  0  0  0  0
0  0  1  1  0  0  0  0
0  0  0  0  1  1  0  0
0  0  0  0  0  0  2  1
0  1  1  0  0  0  0  0
0  0  0  1  1  0  0  0
0  0  0  0  0  1  1  0
0  0  0  0  0  1  1  1

C8 - A7
1  1  0  0  0  0  0  0
0  0  1  1  0  0  0  0
0  0  0  0  1  1  0  0
0  0  0  0  0  0  1  2
0  0  0  0  1  1  0  0
0  0  0  1  1  0  0  0
0  1  1  0  0  0  0  0

C8 - C2
1  4  3  4  5  8  7  6
1  0  2  2  2  0  1  2

C8 - A1
15 28 39 48 55 60 63 64

C8 - A1*D4
1  0  1  0  1  0  1  2
1  2  1  0  0  0  0  0
0  0  1  2  1  0  0  0
0  0  0  0  1  2  1  2
0  0  0  0  1  2  3  2

C8 - A1*C7
0  0  0  0  0  0  0  1
1  0  0  0  0  0  0  0
0  1  0  0  0  0  0  0
0  0  1  0  0  0  0  0
0  0  0  1  0  0  0  0
0  0  0  0  1  0  0  0
0  0  0  0  0  1  0  0
0  0  0  0  0  0  1  1

C8 - C2*C6
0  0  0  0  0  1  1  0
0  0  0  0  0  0  0  1
1  0  0  0  0  0  0  0
0  1  0  0  0  0  0  0
0  0  1  0  0  0  0  0
0  0  0  1  0  0  0  0
0  0  0  0  1  1  0  0
0  0  0  0  0  0  1  1
```

```
C8 - C3*C5
0  0  0  1  1  0  0  0
0  0  0  0  0  1  1  0
0  0  0  0  0  0  0  1
1  0  0  0  0  0  0  0
0  1  0  0  0  0  0  0
0  0  1  1  0  0  0  0
0  0  0  0  1  1  0  0
0  0  0  0  0  0  1  1

C8 - C4*C4
0  1  1  0  0  0  0  0
0  0  0  1  1  0  0  0
0  0  0  0  0  1  1  0
0  0  0  0  0  0  0  1
1  1  0  0  0  0  0  0
0  0  1  1  0  0  0  0
0  0  0  0  1  1  0  0
0  0  0  0  0  0  1  1

D8 - A7
1  1  0  0  0  0  0  0
0  0  1  1  0  0  0  0
0  0  0  0  1  1  0  0
0  0  0  0  0  0  0  1
0  0  0  0  0  1  1  0
0  0  0  1  1  0  0  0
0  1  1  0  0  0  0  0

D8 - B7
1  0  0  0  0  0  0  0
0  1  0  0  0  0  0  0
0  0  1  0  0  0  0  0
0  0  0  1  0  0  0  0
0  0  0  0  1  0  0  0
0  0  0  0  0  1  0  0
0  0  0  0  0  0  1  1

D8 - B4
0  0  0  1  1  0  1  0
0  0  1  0  1  1  0  0
0  1  0  0  0  0  0  1
1  0  1  2  1  2  1  0

D8 - A1*C4
1  2  1  2  1  2  0  1
1  0  1  1  0  0  0  0
0  1  1  0  1  0  0  0
0  0  0  1  1  0  0  1
0  0  0  0  1  1  0  0

D8 - A1*A1*D6
0  0  0  0  1  1  0  0
0  0  0  0  1  1  1  1
1  0  0  0  0  0  0  0
0  1  0  0  0  0  0  0
0  0  1  0  0  0  0  0
0  0  0  1  1  0  0  0
0  0  0  0  0  1  1  0
0  0  0  0  0  1  0  1

D8 - A3*D5
0  0  0  0  1  1  0  0
0  0  1  1  0  0  0  0
0  0  0  0  1  1  1  1
1  0  0  0  0  0  0  0
0  1  1  0  0  0  0  0
0  0  0  1  1  0  0  0
0  0  0  0  0  1  1  0
0  0  0  0  0  1  0  1

D8 - D4*D4
1  1  0  0  0  0  0  0
0  0  1  1  0  0  0  0
0  0  0  0  1  1  1  1
0  1  1  0  0  0  0  0
0  0  0  1  1  0  0  0
0  0  0  0  1  1  0  0
0  0  0  0  0  1  0  1

D8 - A1*B6
0  0  0  0  0  1  1
1  0  0  0  0  0  0
0  1  0  0  0  0  0
0  0  1  0  0  0  0
0  0  0  1  0  0  0
0  0  0  0  1  0  0
0  0  0  0  0  2  1

D8 - C2*B5
0  0  0  0  0  1  1
0  0  0  1  1  0  0
1  0  0  0  0  0  0
0  1  0  0  0  0  0
0  0  1  0  0  0  0
0  0  0  1  1  0  0
0  0  0  0  2  1  1
```

```
D8 - B3*B4
0  0  1  1  0  0  0  0
0  0  0  0  1  1  0  0
0  0  0  0  0  0  1  1
1  0  0  0  0  0  0  0
0  1  1  0  0  0  0  0
0  0  0  1  1  0  0  0
0  0  0  0  0  1  1  0

D8 - C2*C2
1  2  1  2  1  2  2  1
0  0  1  1  2  1  0  1
1  0  1  2  1  2  0  1
0  1  1  0  1  1  1  0

E8 - A4*A4
1  1  1  1  2  2  1  1
0  1  1  1  1  0  0  1
0  0  1  1  1  1  1  0
0  0  0  1  1  0  1  0
0  1  1  1  1  0  0  1
0  0  0  1  1  1  1  1
1  1  1  1  1  0  1  1
0  0  1  1  1  1  1  1

E8 - A1*A2*A5
1  1  1  1  2  2  1  1
0  1  1  1  2  1  1  1
0  0  1  1  2  1  0  1
0  0  1  1  1  1  1  0
1  1  1  1  1  0  0  1
0  1  1  1  1  1  0  0
0  0  1  1  1  1  1  1

E8 - G2*F4
0  0  0  1  1  0  0  1
1  1  2  1  1  2  1  0
0  1  1  1  1  0  0  1
0  0  0  0  1  1  0  1
0  0  1  1  1  1  0  1
1  1  0  1  1  1  0  0

E8 - A1*A7
0  0  0  1  2  1  0  1
0  1  1  1  1  1  1  0
0  0  1  1  1  1  1  0
0  0  1  1  1  0  0  1
1  1  0  0  0  0  0  0
0  0  1  1  1  1  1  0
0  1  1  1  1  0  0  1
0  1  1  1  0  0  0  0

E8 - A2*E6
0  0  1  1  1  1  0  1
0  0  0  1  1  1  1  0
0  0  0  0  1  1  1  1
0  1  1  1  0  0  0  0
0  0  0  1  1  0  0  0
1  0  0  0  0  0  0  0

E8 - A1*E7
0  0  0  0  1  0  0  1
1  0  0  0  0  0  0  0
0  1  0  0  0  0  0  0
0  0  1  1  1  0  0  1
0  0  0  1  0  0  0  1
0  0  0  0  1  1  0  0
0  0  0  1  1  0  0  0

E8 - A3*D5
0  1  1  1  1  1  0  1
0  0  0  1  2  1  1  1
0  0  1  1  1  1  0  1
1  1  1  1  2  1  0  1
0  0  0  0  1  1  1  0
0  0  1  1  1  0  0  1
0  1  1  1  1  0  0  0

E8 - D8
0  0  1  1  1  0  0  1
0  1  1  0  1  1  0  0
0  0  0  1  1  0  0  0
0  0  0  0  1  1  0  1
0  0  1  1  0  0  0  1
0  0  0  0  1  1  0  0
1  1  0  0  0  0  0  0

E8 - A8
0  0  1  1  1  1  1  1
0  0  0  1  1  1  0  0
1  1  1  0  0  0  0  0
0  0  0  1  1  0  0  0
0  0  0  0  1  1  0  0
0  1  1  1  0  0  0  0
0  0  0  1  1  0  0  0
0  0  1  1  1  0  0  0
```

```
E8 - C2
2  8  8 12 16  8  4  6
3  3  7  8  9  8  4  6

E8 - A1*A2
6 10 14 16 22 16  8 12
1  4  4  6  8  4  2  3
1  1  4  6  5  4  2  3

E8 - A1
46  90 132 172 210 142  72 106

E8 - A1
38  74 108 142 174 118  60  88

E8 - A1
58 114 168 220 270 182  92 136
```

Table V

The Lowest Non-Trivial Representations

algebra	ω	dimension
An	(10...0)	n+1
Bn	(10...0)	2n+1
Cn	(10...0)	2n
Dn	(10...0)	2n
E6	(100000)	27
E7	(0000010)	56
E8	(100000000)	248
F4	(0001)	26
G2	(01)	7

Table VI

Projections of ν_i

Algebra	i	Projection $\nu_i F_{ij}$ of vectors ν_i
A_n, C_n	$1,2,\ldots,n$	$\sum\limits_{j=1}^{i} M_j$
Bn	$1,2,\ldots,n-1$	$\sum\limits_{j=1}^{i} M_j$
	n	$\tfrac{1}{2} \sum\limits_{j=1}^{n} M_j$
Dn	$1,2,\ldots,n-2$	$\sum\limits_{j=1}^{i} M_j$
	$n-1$	$\tfrac{1}{2} \sum\limits_{j=1}^{n-1} M_j - \tfrac{1}{2} M_n$
	n	$\tfrac{1}{2} \sum\limits_{j=1}^{n} M_j$
E6	1	M_1
	2	$M_1 + M_2$
	3	$M_1 + M_2 + M_3$
	4	$-M_{26} -M_{27}$

Table VI (continued)

	5	$-M_{27}$
	6	$M_1 + M_2 + M_3 + M_{26} + M_{27}$
E7	1	$M_6 + M_7$
	2	$\frac{1}{2}(M_1 + M_2 + M_3 + M_4 + M_5 + M_7)$
	3	$M_1 + M_2 + M_3 + M_4$
	4	$M_1 + M_2 + M_3$
	5	$M_1 + M_2$
	6	M_1
	7	$M_1 + M_2 + M_3 + M_4 + M_5 - M_7$
E8	1	M_1
	2	$M_1 + M_2$
	3	$M_1 + M_2 + M_3$
	4	$M_1 + M_2 + M_3 + M_4$
	5	$M_1 + M_2 + M_3 + M_4 + M_5$
	6	$\frac{1}{2}(M_1 + M_2 + M_3 + M_4 + M_5 + M_6 + M_8)$
	7	$M_7 + M_8$
	8	$\frac{1}{2}(M_1 + M_2 + M_3 + M_4 + M_5 + M_6 - M_8)$

Table VI   (continued)

F4	1	$M_3 + M_4$
	2	$M_1 + M_2 + M_3$
	3	$M_1 + M_2$
	4	M_1
G2	1	$M_1 + M_2$
	2	M_1

E7
(1000010)

**************** *************** ************** ***************

A7	A1	A1*G2	A1*D6
(0101000)	(61)	(3) (12)	(0) (010001)
(0001010)	(59)	(5) (03)	(1) (000011)
(1100001)	(57)	(1) (12)	(1) (000110)
(1000011)	(55)	2(5) (11)	(2) (100010)
(1000100)	2(53)	2(3) (03)	(1) (001000)
(0010001)	3(51)	(1) (20)	(0) (100010)
(0100000)	3(49)	(7) (02)	(2) (000001)
(0000010)	4(47)	3(3) (11)	(0) (000001)
	5(45)	2(1) (03)	(3) (100000)
E6	6(43)	2(5) (02)	(1) (100000)
(100001)	7(41)	(7) (10)	
(000011)	8(39)	2(1) (11)	A2*A5
2(100010)	9(37)	2(5) (10)	(00) (10101)
(000020)	10(35)	4(3) (02)	(01) (01100)
(200000)	11(33)	(7) (01)	(10) (00110)
(000100)	12(31)	3(3) (10)	(10) (20001)
(010000)	13(29)	3(1) (02)	(11) (11000)
2(000001)	14(27)	3(5) (01)	(20) (10010)
3(000010)	14(25)	3(1) (10)	(01) (10002)
3(100000)	14(23)	4(3) (01)	(02) (01001)
2(000000)	14(21)	3(1) (01)	(11) (00011)
	15(19)	(5) (00)	2(01) (10010)
A2	14(17)	(3) (00)	2(10) (01001)
(10, 4)	13(15)		(00) (11000)
(4,10)	12(13)	A1*A1	(00) (00011)
(8, 5)	11(11)	(14) (5)	2(11) (00100)
(5, 8)	10(9)	(12) (7)	(21) (10000)
(9, 3)	8(7)	(10) (9)	(12) (00001)
2(6, 6)	6(5)	(14) (3)	(02) (10000)
(3, 9)	4(3)	2(12) (5)	(20) (00001)
(7, 4)	2(1)	3(10) (7)	2(00) (00100)
(4, 7)		2(8) (9)	2(10) (10000)
2(8, 2)	A1*F4	(6) (11)	2(01) (00001)
2(5, 5)	(1) (1001)	(14) (1)	
2(2, 8)	(3) (0002)	3(12) (3)	A1*A3*A3
2(6, 3)	(3) (0010)	5(10) (5)	(0) (110) (110)
2(3, 6)	(1) (0002)	5(8) (7)	(0) (011) (011)
2(7, 1)	(1) (0010)	3(6) (9)	(1) (200) (101)
2(4, 4)	(3) (1000)	(4) (11)	(1) (101) (200)
2(1, 7)	(1) (1000)	6(10) (3)	(1) (101) (002)
2(5, 2)	(5) (0001)	8(8) (5)	(1) (020) (010)
2(2, 5)	2(3) (0001)	2(12) (1)	(1) (002) (101)
2(3, 3)	2(1) (0001)	7(6) (7)	(1) (010) (020)
2(6, 0)	(5) (0000)	3(4) (9)	(0) (201) (100)
2(0, 6)	(3) (0000)	(2) (11)	(2) (110) (001)
2(4, 1)	(1) (0000)	4(10) (1)	(0) (102) (001)
2(1, 4)		9(8) (3)	(2) (001) (100)
2(2, 2)	G2*C3	11(6) (5)	(2) (100) (011)
	(02) (110)	7(4) (7)	(0) (100) (201)
A1	(10) (110)	3(2) (9)	(2) (001) (110)
(47)	(01) (011)	6(8) (1)	(0) (001) (102)
(45)	(02) (001)	12(6) (3)	(1) (111) (000)
2(43)	(11) (100)	11(4) (5)	3(1) (101) (010)
3(41)	(01) (300)	6(2) (7)	3(1) (010) (101)
4(39)	(10) (001)	(0) (9)	(1) (000) (111)
6(37)	2(01) (110)	8(6) (1)	2(0) (110) (001)
7(35)	(00) (201)	12(4) (3)	2(0) (100) (011)
10(33)	(02) (100)	9(2) (5)	2(0) (011) (100)
12(31)	(10) (100)	2(0) (7)	2(0) (001) (110)
14(29)	(01) (001)	8(4) (1)	2(2) (100) (100)
17(27)	(00) (110)	9(2) (3)	2(2) (001) (001)
19(25)	2(01) (100)	3(0) (5)	(1) (200) (000)
21(23)	(00) (001)	6(2) (1)	(3) (010) (000)
23(21)	(00) (100)	3(0) (3)	(1) (002) (000)
24(19)		4(0) (0)	(3) (000) (010)
25(17)			(1) (000) (200)
25(15)			(1) (000) (002)
24(13)			3(0) (100) (100)
23(11)			3(0) (001) (001)
20(9)			3(1) (010) (000)
17(7)			3(1) (000) (010)
14(5)			
9(3)			
10(0)			

Table VII

Branching rules for the representation (1000010) of E_7
with respect to all maximal semisimple subalgebras of E_7.

```
      PROGRAM BRULES(INPUT,OUTPUT)                                    10
      INTEGER Z,ZRANK,HWZ,R,A,ADJN                                    20
      INTEGER W,WADJ,ALGNAM,CHECK                                     30
      INTEGER SNAM,SRANK,SSTAR,FIJ,HRANK                              40
      COMMON /COM0/ ALGNAM(7),CHECK                                   50
      COMMON /COM1/ HWZ(8),Z,ZRANK,NDIMZ,SUM2Z                        60
      COMMON /COM2/ A(8,8),V(8,8),R(8),ADJN(8)                        70
      COMMON /COM3/ WADJ(130),LVADJ(130),NWADJ,NWAMAX                 80
      COMMON /COM4/ W(5000),LV(5000),MU(5000),NW,NWMAX                90
      COMMON /COM5/ SNAM(8),SRANK(8),SSTAR(8),FIJ(8,8),NSUB,HRANK    100
C                                                                    110
$--THIS IS THE MAIN PROGRAM FOR                                     120
$--'THE COMPUTATION OF BRANCHING RULES FOR REPRESENTATIONS OF SEMISIMPLE  130
$   LIE ALGEBRAS' - BY W.MCKAY, J.PATERA, D.SANKOFF.               140
$--THIS VERSION CALCULATES THE BRANCHING RULES FOR A GIVEN HIGHEST  150
$--WEIGHT OF A SIMPLE ALGEBRA Z AND A SEMISIMPLE SUBALGEBRA H.      160
C                                                                    170
$--DESCRIPTION OF COMMON VARIABLES                                  180
$--ALGNAM - CONTAINS THE LETTERS A,B,C,D,E,F,G FOR DECODING THE ALGEBRA NAME.  190
$--CHECK  - IS AN INPUT PARAMETER =1 IF N,I AND Q ARE TO BE OUTPUT. OTHERWISE =0  200
$--HWZ    - CONTAINS THE HIGHEST WEIGHT OF THE REPRESENTATION OF Z.  210
$--Z      - CONTAINS THE CODE CORRESPONDING TO THE ALGEBRA NAME.    220
$--ZRANK  - IS THE RANK OF Z.                                       230
$--NDIMZ  - IS THE DIMENSION OF THE REPRESENTATION OF Z.            240
$--SUM2Z  - IS THE 2ND ORDER INDEX OF THE REPRESENTATION OF Z.      250
$--A      - IS THE CARTAN MATRIX.                                   260
$--V      - IS THE SCALAR PRODUCT MATRIX.                           270
$--R      - IS THE LEVEL VECTOR.                                    280
$--ADJN   - IS THE ADJOINT REPRESENTATIONO OF Z.                    290
$--WADJ   - CONTAINS THE POSITIVE ROOTS.                            300
$--LVADJ  - CONTAINS THE LEVELS OF THE POSITIVE ROOTS.              310
$--NWADJ  - IS THE NUMBER OF POSITIVE ROOTS.                        320
$--NWAMAX - IS THE MAXIMUM NUMBER OF POSITIVE ROOTS ALLOWED (=130). 330
$--W      - CONTAINS THE WEIGHTS IN THE RESTRICTED WEIGHT SYSTEM.   340
$--LV     - CONTAINS THE LEVEL OF THE WEIGHTS.                      350
$--MU     - CONTAINS THE MULTIPLICITY OF THE WEIGHTS.               360
$--NW     - IS THE NUMBER OF WEIGHTS.                               370
$--NWMAX  - IS THE MAXIMUM NUMBER OF WEIGHTS ALLOWED (=5000).       380
$--SNAM   - CONTAINS THE NAME CODES OF THE SIMPLE SUBALGEBRA COMPONENTS  390
$--       - OF THE NONSIMPLE SUBALGEBRA H (=H1*H2*...).             400
$--SRANK  - CONTAINS THE RANKS OF THE COMPONENTS OF H.              410
$--SSTAR  - CONTAINS THE SEPARATOR * OF THE COMPONENTS OF H.        420
$--FIJ    - IS THE PROJECTION MATRIX OF (Z-H).                      430
$--NSUB   - IS THE NUMBER OF SIMPLE SUBALGEBRA COMPONENTS OF H.     440
$--HRANK  - IS THE RANK OF H.                                       450
      CALL DATAZ                                                     460
      CALL WTSYSZ                                                    470
      CALL DATAH                                                     480
      CALL PROJECT                                                   490
      STOP 0                                                         500
      END                                                            510
      BLOCKDATA                                                      520
      INTEGER ALGNAM,CHECK                                           530
      COMMON /COM0/ ALGNAM(7),CHECK                                  540
      COMMON /COM3/ WADJ(130),LVADJ(130),NWADJ,NWAMAX               550
      COMMON /COM4/ W(5000),LV(5000),MU(5000),NW,NWMAX              560
      DATA ALGNAM/1HA,1HB,1HC,1HD,1HE,1HF,1HG/,CHECK/0/             570
      DATA WADJ/130*0/,LVADJ/130*1/,NWAMAX/130/                     580
      DATA W/5000*0/,LV/5000*0/,MU/5000*0/,NWMAX/5000/             590
      END                                                            600
      SUBROUTINE CALCWT(WTS,LEV,HW,AA,M,LOWLV,NWTS,NWTSMX)          610
      INTEGER WTS(1),LEV(1),HW(8),AA(8,8)                           620
      INTEGER XN(8),WN(8),X                                         630
$--THIS SUBROUTINE COMPUTES THE WEIGHTS WTS AND LEVELS LEV STARTING  640
$--FROM A HIGHEST WEIGHT HW (ASSUMED TO BE AT LEVEL 1) UP TO THE LEVEL LOWLV  650
$--AA IS THE CARTAN MATRIX OF THE ALGEBRA Z WITH HIGHEST WEIGHT HW  660
$--M  IS THE RANK OF Z                                              670
$--WN + XN ARE WORKING ARRAYS WHICH STORE THE WEIGHTS IN UNPACKED FORM  680
$--IN WTS THE WEIGHTS ARE IN PACKED FORM.                           690
$--THE METHOD IS AS FOLLOWS:                                        700
$--STARTING FROM THE HIGHEST WEIGHT WITH LEVEL=1, THE NEW WEIGHTS IN THE  710
$--SYSTEM ARE FOUND BY SUBTRACTING THE K SIMPLE ROOTS FROM EACH     720
$--POSITIVE (.GE. 1) CO-ORDINATE K OF THE EXISTING WEIGHTS IN WTS.  730
$--ONLY DISTINCT WEIGHTS ARE KEPT.  THIS METHOD DOES NOT ENSURE THAT  740
$--THE WEIGHTS ARE GENERATED IN ORDER OF LEVELS.                    750
      CALL PACK(HW,M,WTS(1))                                        760
      LEV(1)=I=NWTS=1                                               770
   10 IF(I.GT.NWTS) RETURN                                          780
      IF(LEV(I).GE.LOWLV) GO TO 70                                  790
      CALL UNPACK (WN,M,WTS(I))                                     800
```

```
      DO 60 J=1,M                                                        810
C--SKIP IF COMPONENT OF WEIGHT HAS A NEGATIVE VALUE....                  820
      IF(WN(J).LT.1) GO TO 60                                           830
      NS=WN(J)                                                          840
      DO 50 K=1,NS                                                      850
      IZ=LEV(I)+K                                                       860
      IF(IZ.GT.LOWLV) GO TO 50                                         870
      DO 20 JJ=1,M                                                      880
   20 XN(JJ)=WN(JJ)-K*AA(J,JJ)                                          890
      CALL PACK(XN,M,X)                                                 900
C--CHECK IF WEIGHT IS ALREADY THERE                                     910
      DO 30 LL=1,NWTS                                                   920
      IF(X.EQ.WTS(LL)) GO TO 50                                        930
   30 CONTINUE                                                          940
      NWTS=NWTS+1                                                       950
      IF(NWTS.GT.NWTSMX) CALL ERR(3,NWTSMX)                             960
      WTS(NWTS)=X                                                       970
      LEV(NWTS)=IZ                                                      980
   50 CONTINUE                                                          990
$
$
   60 CONTINUE                                                          1000
   70 I=I+1                                                             1010
      GO TO 10                                                          1020
      END                                                               1030
      SUBROUTINE DATAH                                                  1040
      INTEGER ASTAR,SNAM,SRANK,SSTAR,FIJ,HRANK                          1050
      INTEGER ALGNAM,CHECK,Z,ZRANK,HWZ                                  1060
      COMMON /COM0/ ALGNAM(7),CHECK                                     1070
      COMMON /COM1/ HWZ(8),Z,ZRANK,NDIMZ,SUM2Z                          1080
      COMMON /COM5/ SNAM(8),SRANK(8),SSTAR(8),FIJ(8,8),NSUB,HRANK       1090
      DATA ASTAR/1H*/                                                   1100
C--INPUT NON-SIMPLE SUBALGEBRA H AND ITS PROJECTION MATRIX FIJ          1110
C--H MAY BE OF THE FORM H1*H2*H3...                                     1120
C--DETERMINE THE NUMBER OF COMPONENTS NSUB                              1130
      READ 1000, (SNAM(J),SRANK(J),SSTAR(J),J=1,8)                      1140
      NSUB=0                                                            1150
      HRANK=0                                                           1160
      IF(SRANK(1).EQ.0) RETURN                                          1170
      DO 10 KK=1,8                                                      1180
      HRANK=HRANK+SRANK(KK)                                             1190
      NSUB=NSUB+1                                                       1200
      IF(SSTAR(KK).NE.ASTAR) GO TO 20                                   1210
   10 CONTINUE                                                          1220
   20 PRINT 2000,(SNAM(J),SRANK(J),SSTAR(J),J=1,NSUB)                   1230
      DO 29 I=1,NSUB                                                    1240
      IHNAM=SNAM(I)                                                     1250
      DO 25 J=1,7                                                       1260
      JH=J                                                              1270
      IF(ALGNAM(J).EQ.IHNAM) GO TO 27                                   1280
   25 CONTINUE                                                          1290
      GO TO 29                                                          1300
   27 SNAM(I)=JH                                                        1310
   29 CONTINUE                                                          1320
      DO 30 I=1,HRANK                                                   1330
      READ 1010, (FIJ(I,J),J=1,ZRANK)                                   1340
   30 CONTINUE                                                          1350
      RETURN                                                            1360
 1000 FORMAT(8(A1,I1,A1))                                               1370
 1010 FORMAT(8I3)                                                       1380
 2000 FORMAT(*0*6X,8(A1,I1,A1))                                         1390
      END                                                               1400
      SUBROUTINE DATAZ                                                  1410
      INTEGER ALGNAM,CHECK,ZNAM                                         1420
      INTEGER Z,ZRANK,HWZ                                               1430
      COMMON /COM0/ ALGNAM(7),CHECK                                     1440
      COMMON /COM1/ HWZ(8),Z,ZRANK,NDIMZ,SUM2Z                          1450
$--INPUT THE SIMPLE ALGEBRA NAME ZNAM OF RANK ZRANK AND HIGHEST WEIGHT HWZ.  1460
$--THE SIMPLE LIE ALGEBRAS DEFINED IN THIS PROGRAM ARE                  1470
$--(A1,...,A8),(B2,...,B8),(C2,...,C8),(D4,...,D8),E6,E7,E8,F4,G2.      1480
$--ASSIGN A CODE Z=(1,2,...,7) CORRESPONDING TO ZNAM=(A,B,...,G)        1490
      PRINT 1020                                                        1500
      READ 1010,CHECK                                                   1510
      READ 1000, ZNAM,ZRANK,(HWZ(J),J=1,ZRANK)                          1520
      IF(ZRANK.LE.0 .OR. ZRANK.GT.8                                     1530
     1 .OR. (Z.EQ.2 .AND. ZRANK.LT.3) .OR. (Z.EQ.3 .AND. ZRANK.LT.2)    1540
     2 .OR. (Z.EQ.4 .AND. ZRANK.LT.4)) GO TO 30                         1550
C--STORE EQUIVALENT INTEGER VALUE IN Z FOR THE CORRESPONDING ALGEBRA NAME  1560
      DO 20 I=1,7                                                       1570
      Z=I                                                               1580
      IF(ALGNAM(I).EQ.ZNAM) GO TO 70                                    1590
   20 CONTINUE                                                          1600
```

```
      30 PRINT 1050, ZNAM,ZRANK                                      1610
         CALL ERR(1,0)                                               1620
      70 PRINT 1030, ZNAM,ZRANK                                      1630
         IF(CHECK.GE.1) PRINT 1035                                   1640
         PRINT 1040,(HWZ(J),J=1,ZRANK)                               1650
         RETURN                                                      1660
    1000 FORMAT(A1,I1,1X,8I3)                                        1670
    1010 FORMAT(I1)                                                  1680
    1020 FORMAT(*1*)                                                 1690
    1030 FORMAT(//*0*6X,A1,I1)                                       1700
    1035 FORMAT(*+*35X*N*15X*I*11X*Q*)                               1710
    1040 FORMAT(7X,8I3)                                              1720
    1050 FORMAT(*0...Z = *A1,I1)                                     1730
         END                                                        1740
         SUBROUTINE ERR(I,J)                                         1750
$--THERE ARE 3 ERRORS/RESTRICTIONS THAT WILL STOP THE PROGRAM        1760
$--ABNORMALLY WITH AN APPROPRIATE MESSAGE.                           1770
$--1) IF THE GIVEN ALGEBRA Z IS UNDEFINED.                           1780
$--2) IF THE COORDINATES (C1,C2,...,CN) OF A WEIGHT DO NOT HAVE      1790
$--    VALUES IN THE RANGE (-50,50)                                  1800
$--3) IF THE NUMBER OF WEIGHTS GENRATED IN A RESTRICTED WEIGHT       1810
$--    SYSTEM EXCEEDS 5000. (OR EXCEEDS 130 - IN THE CASE OF         1820
$--    A ROOT SYSTEM)                                                1830
         IF(I-2) 1,2,3                                               1840
       1 PRINT 100                                                   1850
         GO TO 90                                                    1860
       2 PRINT 200                                                   1870
         GO TO 90                                                    1880
       3 PRINT 300,J                                                 1890
      90 STOP 77                                                     1900
     100 FORMAT(*0...CALLED FROM DATAZ - Z IS UNDEFINED IN THIS PROGRAM.*  1910
        1       * SEE SECTION II, METHOD (A).*)                      1920
     200 FORMAT(*0...CALLED FROM PACK - COORDINATES OF A WEIGHT TO BE*  1930
        1       * PACKED ARE NOT ALL IN THE RANGE (-50,50).*/        1940
        2       * SEE SECTION III, THE PROGRAM.*)                    1950
     300 FORMAT(*0...CALLED FROM CALCWT - THE NUMBER OF WEIGHTS GENERATED*  1960
        1       * EXCEEDS THE MAXIMUM ALLOWED =*I5)                  1970
         END                                                        1980
         SUBROUTINE FREUD(WTS,LEV,MUL,ROOTS,C,NWTS,NROOTS,M,MUHW,NDIM,  1990
$
$
        1              LOWLV,TEVEN,SUM2)                             2000
         LOGICAL TEVEN                                               2010
         INTEGER WTS(1),ROOTS(1),LEV(1),MUL(1)                       2020
         INTEGER WN(8),XN(8),BN(8),UN(8),U                          2030
         DIMENSION C(8,8)                                            2040
$--THIS SUBROUTINE CALCULATES THE MULTIPLICITIES MUL OF THE WEIGHTS  2050
$--IN THE RESTRICTED WEIGHT SYSTEM (WTS,LEV) USING FREUDENTHALS      2060
$--RECURSION FORMULA,.  IT ALSO CALCULATES THE DIMENSION NDIM AND THE  2070
$--INDEX SUMM2 OF THE I.R. FOR THIS SYSTEM BY COUNTING MULTIPLICITIES.  2080
$--FIRST WEIGHT HAS MULTIPLICITY 1                                   2090
         CALL UNPACK(WN,M,WTS(1))                                    2100
         MUL(1)=1                                                    2110
         NDIM=SUM2=0                                                 2120
         IDUP=2 $ IF(TEVEN .AND.LEV(1).EQ.LOWLV) IDUP=1              2130
         NDIM=NDIM+IDUP                                              2140
         CALL SPRD(C,WN,WN,M,S2)                                     2150
         SUM2=SUM2+S2*IDUP                                           2160
         IF(NWTS.LE.1) GO TO 500                                     2170
$--FIND PART OF LHS OF FORMULA.  N.B. DELTA=(1,1,...,1)              2180
         DO 170 J=1,M                                                2190
     170 XN(J)=WN(J)+1                                               2200
         CALL SPRD(C,XN,XN,M,S1)                                     2210
         DO 400 I=2,NWTS                                             2220
         CALL UNPACK(WN,M,WTS(I))                                    2230
C--FIND REST OF LHS FOR CURRENT WT                                   2240
         DO 180 J=1,M                                                2250
     180 XN(J)=WN(J)+1                                               2260
         CALL SPRD(C,XN,XN,M,S2)                                     2270
         SLHS=S1-S2                                                  2280
         SRHS=0                                                      2290
         DO 310 LA=1,NROOTS                                          2300
         IF(ROOTS(LA).EQ.0)GO TO 310                                 2310
         CALL UNPACK(BN,M,ROOTS(LA))                                 2320
         KS=0                                                        2330
     200 KS=KS+1                                                     2340
         DO 260 J=1,M                                                2350
     260 UN(J)=WN(J)+KS*BN(J)                                        2360
         CALL PACK(UN,M,U)                                           2370
C--LOOK FOR THIS WEIGHT IN THE SYSTEM                                2380
```

```
      DO 290 L=1,NWTS                                          2390
      IF(U.NE.WTS(L)) GO TO 290                                2400
      CALL SPRD(C,BN,UN,M,S)                                   2410
      X=MUL(L)                                                 2420
      SRHS=SRHS+2.*S*X                                         2430
      GO TO 200                                                2440
  290 CONTINUE                                                 2450
  310 CONTINUE                                                 2460
  320 MUL(I)=SRHS/SLHS+.0001                                   2470
      IDUP=2 $ IF(TEVEN .AND. LEV(I).EQ.LOWLV) IDUP=1          2480
      IF(LEV(I).GT.LOWLV) IDUP=0                               2490
      NDIM=NDIM+MUL(I)*IDUP                                    2500
      CALL SPRD(C,WN,WN,M,S2)                                  2510
      SUM2=SUM2+S2*MUL(I)*IDUP                                 2520
  400 CONTINUE                                                 2530
  500 NDIM=NDIM*MUHW                                           2540
      SUM2=SUM2*MUHW                                           2550
      RETURN                                                   2560
      END                                                      2570
      SUBROUTINE GETLL(HW,R,M,LOWLV,T)                         2580
      INTEGER HW(8),R(8),T                                     2590
C--THIS SUBROUTINE FINDS THE LOWEST LEVEL LOWLV OF THE RESTRICTED   2610   2600
C--WEIGHT SYSTEM OF A SIMPLE ALGEBRA WITH HIGHEST WEIGHT HW,        2610
C--AND THE NUMBER OF GAPS T IN THE SYSTEM.                          2620
      T=0                                                      2630
      DO 100 J=1,M                                             2640
  100 T=T+R(J)*HW(J)                                           2650
      LOWLV=(T+2)/2                                            2660
      RETURN                                                   2670
      END                                                      2680
      SUBROUTINE GETLLH(HW,RH,NAMES,RANKS,NS,LLH,TH)           2690
      INTEGER HW(8),RH(8),NAMES(8),RANKS(8),TH                 2700
      INTEGER WN(8),R(8),H,T                                   2710
C--FIND THE LOWEST LEVEL LLH OF THE NON-SIMPLE SUBALGEBRA H=H1*H2*...  2720
$--LLH=(T(H1),T(H2)+...)/2+1                                   2730
$--AND THE NUMBER OF GAPS TH                                   2740
      LLH=0                                                    2750
      NLAST=0                                                  2760
      DO 50 KK=1,NS                                            2770
      H=NAMES(KK)                                              2780
      NH=RANKS(KK)                                             2790
      DO 10 J=1,NH                                             2800
      NLJ=NLAST+J                                              2810
      WN(J)=HW(NLJ)                                            2820
      R(J)=RH(NLJ)                                             2830
   10 CONTINUE                                                 2840
      CALL GETLL(WN,R,NH,LL,T)                                 2850
      TH=LLH=LLH+T                                             2860
      NLAST=NLAST+NH                                           2870
   50 CONTINUE                                                 2880
      LLH=(LLH/2)+1                                            2890
      RETURN                                                   2900
      END                                                      2910
      SUBROUTINE PACK(WN,M,WT)                                 2920
      INTEGER WN,WT,S,T                                        2930
      DIMENSION WN(8)                                          2940
      DATA I15/100000000000000/,I16/1000000000000000/         2950
C--THIS SUBROUTINE PACKS THE CONTENTS OF ARRAY WN CONTAINING A STRING  2960
C--OF UP TO 8 2-DIGIT NUMBERS IN THE RANGE -50 TO 50          2970
C--INTO A VARIABLE WT                                          2980
C--WN MAY HAVE NEGATIVE COMPONENTS AND IS RESTRICTED TO THE ABOVE RANGE  2990
$
$
C--SO THAT BY ADDING 50 TO EACH NUMBER THE RANGE WILL NOT EXCEED 99   3000
C--THIS ALLOWS THE 8 NUMBERS TO BE PACKED INTO A SINGLE VARIABLE ON THE  3010
C--CDC MACHINES.  AN INTEGER CONSTANT STRING MAY CONTAIN UP TO 15     3020
$--DECIMAL DIGITS.  HOWEVER IF DIGITS 15 AND 16 ARE TREATED           3030
$--SEPARATELY WE MAY PACK 16 DIGITS INTO WT.                          3040
C--THE METHOD IS TO PACK THE 8 NUMBERS IN WN AS A 16 DIGIT DECIMAL NUMBER  3050
C--SO THAT THE DIGITS WN(J) ARE IN THE POSITIONS (2J-2) AND (2J-1) OF W.  3060
C--E.G.  WN(J)= 11 22 33 44 55 66 77 88                       3070
C--THEN WT= 887766554432211                                   3080
C--IF M=1 NO PACKING IS DONE                                  3090
      IF(M.NE.1) GO TO 20                                      3100
      WT=WN(1)                                                 3110
      RETURN                                                   3120
C--ADD 50 TO VALUE OF WN TO MAKE THEM NON-NEGATIVE            3130
   20 DO 30 I=1,M                                              3140
      WN(I)=50+WN(I)                                           3150
      IF(WN(I).GE.0 .AND. WN(I).LE.99) GO TO 30               3160
```

```
        CALL ERR(2,0)                                               3170
   30 CONTINUE                                                      3180
        WT=0                                                        3190
   40 IF(M.LT.8) GO TO 80                                           3200
C--PACK 1ST DIGIT OF WN(8) AS 15TH DIGIT OF W                       3210
        S=10*(WN(8)/10)                                             3220
        T=WN(8)-S                                                   3230
        IF(T.EQ.0) GO TO 60                                         3240
        DO 50 I=1,T                                                 3250
   50 WT=WT+I15                                                     3260
C--PACK 2ND DIGIT OF WN(8) AS 16TH DIGIT OF W                       3270
   60 T=(WN(8)-T)/10                                                3280
        IF(T.EQ.0) GO TO 80                                         3290
        DO 70 I=1,T                                                 3300
   70 WT=WT+I16                                                     3310
C--PACK REMAINING DIGITS                                            3320
   80 MM=7                                                          3330
        IF(M.LT.7) MM=M                                             3340
        DO 90 J=1,MM                                                3350
C--PACK THE 1ST DIGIT OF WN(J) IN POSITION (2J-2)                   3360
        S=10*(WN(J)/10)                                             3370
        T=WN(J)-S                                                   3380
        WT=WT+T*10**(2*J-2)                                         3390
C--PACK THE 2ND DIGIT OF WN(J) IN POSITION (2J-1)                   3400
        T=(WN(J)-T)/10                                              3410
   90 WT=WT+T*10**(2*J-1)                                           3420
C--RESTORE ORIGINAL VALUES OF WN                                    3430
        DO 100 I=1,M                                                3440
  100 WN(I)=WN(I)-50                                                3450
        RETURN                                                      3460
        END                                                         3470
        SUBROUTINE PROJECT                                          3480
        LOGICAL TEVEN                                               3490
        INTEGER ALGNAM,CHECK                                        3500
        INTEGER Z,ZRANK,HWZ,R,A,ADJN,W,WADJ                         3510
        INTEGER SNAM,SRANK,SSTAR,FIJ,HRANK,H,T,TH,T1                3520
        INTEGER PW(5000),MUPW(5000),PWN(8),ADJNH(8),RH(8),WNH(8),   3530
     1        WN(8),AH(8,8),PWI                                     3540
        DIMENSION VH(8,8)                                           3550
        COMMON /COM0/ ALGNAM(7),CHECK                               3560
        COMMON /COM1/ HWZ(8),Z,ZRANK,NDIMZ,SUM2Z                    3570
        COMMON /COM2/ A(8,8),V(8,8),R(8),ADJN(8)                    3580
        COMMON /COM3/ WADJ(130),LVADJ(130),NWADJ,NWAMAX             3590
        COMMON /COM4/ W(5000),LV(5000),MU(5000),NW,NWMAX            3600
        COMMON /COM5/ SNAM(8),SRANK(8),SSTAR(8),FIJ(8,8),NSUB,HRANK 3610
$--THIS SUBROUTINE FINDS THE BRANCHING RULES FOR THE HWZ            3620
$--REPRESENTATION OF (Z-H) AS FOLLOWS:                              3630
$--(A)..THE RESTRICTED WEIGHT SYSTEM (W,LV,MU) IS PROJECTED         3640
$--INTO A RESTRICTED WEIGHT SYSTEM (PW,MUPW) BY APPLYING THE        3650
$--PROJECTION MATRIX FIJ TO EVERY WEIGHT OF W.  ONLY WEIGHTS WITH   3660
$--NON-NEGATIVE COORDINATES ARE KEPT.                               3670
$--(B).. THE CARTAN MATRIX AH, THE SCALAR PRODUCT MATRIX WH, THE    3680
$--LEVEL VECTOR RH AND ADJOINT REPRESENATION ADJNH OF THE NONSIMPLE 3690
$--H ARE FOUND.                                                     3700
$--(C).. THE LEVELS OF THE WEIGHTS IN (PW,MUPW) ARE CALCULATED AND SORTED. 3710
$--(D).. THE IRREDUCIBLE COMPONENTS  IN (PW,MUPW) ARE SELECTED OUT  3720
$--STARTING FROM THE TOP LEVEL(=1).  AS EACH IRREDUCIBLE COMPONENT IS 3730
$--FOUND, ITS WEIGHT SYSTEM IS GENERATED AND SUBTRACTED FROM THE    3740
$--REMAINING SYSTEM(PW,MUPW) BY SUBTRACTING MULTIPLICITES.  THE NEXT 3750
$--WEIGHT WITH NON-ZERO MULTIPLICITY WILL BE A HIGHEST WEIGHT SINCE 3760
$--THE SYSTEM (PW,MUPW) IS ORDERED BY LEVELS.                       3770
C                                                                   3780
$--(A)                                                              3790
   25 NPW=0                                                         3800
        DO 40 I=1,NW                                                3810
        CALL UNPACK(WN,ZRANK,W(I))                                  3820
        DO 32 J=1,HRANK                                             3830
        PWN(J)=0                                                    3840
        DO 30 K=1,ZRANK                                             3850
   30 PWN(J)=PWN(J)+FIJ(J,K)*WN(K)                                  3860
$--ELIMINATE THIS PROJECTED WEIGHT IF ANY OF ITS COMPONENT IS       3870
$--NON-NEGATIVE BECAUSE IT CANNOT BE A HIGHEST WEIGHT.              3880
        IF(PWN(J).LT.0) GO TO 40                                    3890
   32 CONTINUE                                                      3900
        MUI=MU(I)                                                   3910
        CALL PACK(PWN,HRANK,PWI)                                    3920
        IF(I.EQ.1) GO TO 37                                         3930
C--CHECK IF THIS PROJECTED WEIGHT ALREADY OCCURS                    3940
        DO 35 IL=1,NPW                                              3950
        IF(PW(IL).NE.PWI) GO TO 35                                  3960
```

```
       MUPW(IL)=MUPW(IL)+MUI                                          3970
       GO TO 40                                                      3980
   35 CONTINUE                                                       3990
$
$
   37 NPW=NPW+1                                                      4000
      PW(NPW)=PWI                                                    4010
      MUPW(NPW)=MUI                                                  4020
   40 CONTINUE                                                       4030
$--(B)                                                               4040
      DO 45 I=1,8                                                    4050
      RH(I)=ADJNH(I)=0                                               4060
      DO 45 J=1,8                                                    4070
   45 AH(I,J)=VH(I,J)=0                                              4080
      NLAST=NWADJ=0                                                  4090
      DO 60 KK=1,NSUB                                                4100
      H=SNAM(KK)                                                     4110
      NH=SRANK(KK)                                                   4120
      CALL SETUP(H,NH,A,V,R,ADJN)                                    4130
$--CALCULATE THE WEIGHT SYSTEM OF THE ADJOINT REPN OF THE CURRENT    4140
$--SUBALGEBRA AND SAVE AS POSITIVE ROOTS OF H                        4150
$--WITH REMAINING COMPONENTS ZERO                                    4160
      CALL GETLL(ADJN,R,NH,LLADJ,T)                                  4170
      LLADJ=LLADJ-1                                                  4180
      CALL CALCWT(WH,LVADJ,ADJN,A,NH,LLADJ,NWH,NWAMAX)               4190
      DO 49 J=1,NWH                                                  4200
      CALL UNPACK(WNH,NH,WH(J))                                      4210
      DO 47 I=1,8                                                    4220
   47 WN(I)=0                                                        4230
      DO 48 I=1,NH                                                   4240
   48 WN(NLAST+I)=WNH(I)                                             4250
      NWADJ=NWADJ+1                                                  4260
      CALL PACK(WN,HRANK,WADJ(NWADJ))                                4270
   49 CONTINUE                                                       4280
C--COPY A,V INTO THE CORRESPONDING BLOCKS OF AH,VH,                  4290
      DO 50 I=1,NH                                                   4300
      NLI=NLAST+I                                                    4310
      RH(NLI)=R(I)                                                   4320
      ADJNH(NLI)=ADJN(I)                                             4330
      DO 50 J=1,NH                                                   4340
      NLJ=NLAST+J                                                    4350
      VH(NLI,NLJ)=V(I,J)                                             4360
      AH(NLI,NLJ)=A(I,J)                                             4370
   50 CONTINUE                                                       4380
      NLAST=NLAST+NH                                                 4390
   60 CONTINUE                                                       4400
$--(C)                                                               4410
$--DETERMINE THE LEVELS OF THE PROJECTED WEIGHT SYSTEM AND ORDER ACCORDING  4420
$--TO THE RELATIVE T VALUES CALCULATED USING GETLL                  4430
      CALL UNPACK(PWN,HRANK,PW(1))                                   4440
      CALL GETLLH(PWN,RH,SNAM,SRANK,NSUB,LLDUM,T1)                   4450
      LV(1)=1                                                        4460
      DO 61 I=2,NPW                                                  4470
      CALL UNPACK(PWN,HRANK,PW(I))                                   4480
      CALL GETLLH(PWN,RH,SNAM,SRANK,NSUB,LLDUM,T)                    4490
   61 LV(I)=T1-T+1                                                   4500
      CALL SORT2(LV,PW,MUPW,NPW)                                     4510
$--(D)                                                               4520
C--TAKING EACH WEIGHT IN PW OF NON-ZERO MULTIPLICITY IN TURN         4530
C--FIND THE CORRESPONDING WEIGHT SYSTEM                              4540
   62 DO 63 I=1,NW                                                   4550
   63 W(I)=LV(I)=MU(I)=0                                             4560
      NW=NDIMH=SUM2H=0                                               4570
      DO 100 I=1,NPW                                                 4580
      IF(NDIMZ-NDIMH.LE.0) GO TO 103                                 4590
      IF(MUPW(I).LE.0) GO TO 100                                     4600
      MPWI=MUPW(I)                                                   4610
      CALL UNPACK(PWN,HRANK,PW(I))                                   4620
C--GET LOWEST LEVEL FOR THIS WEIGHT                                  4630
      CALL GETLLH(PWN,RH,SNAM,SRANK,NSUB,LLH,TH)                     4640
      CALL CALCWT(W,LV,PWN,AH,HRANK,LLH,NW,NWMAX)                    4650
      CALL SORT1(LV,W,NW)                                            4660
      TEVEN=.FALSE.                                                  4670
      IF(MOD(TH,2).EQ.0) TEVEN=.TRUE.                                4680
      CALL FREUD(W,LV,MU,WADJ,VH,NW,NWADJ,HRANK,MPWI,NDIM,LLH,TEVEN, 4690
     1SUM2)                                                          4700
$--SUBTRACT WEIGHTS OUT OF THE PROJECTED SYSTEM                      4710
      DO 70 IL=1,NW                                                  4720
      DO 65 IS=1,NPW                                                 4730
      IF(W(IL).NE.PW(IS)) GO TO 65                                   4740
```

```
      MUPW(IS)=MUPW(IS)-MPWI*MU(IL)                                      4750
      GO TO 70                                                          4760
   65 CONTINUE                                                          4770
   70 CONTINUE                                                          4780
      IF(MPWI.EQ.1) PRINT 2020,(PWN(J),J=1,HRANK)                       4790
      IF(MPWI.GT.1) PRINT 2010,MPWI,(PWN(J),J=1,HRANK)                  4800
      IS2=SUM2+0.005                                                    4810
      IF(CHECK.GE.1) PRINT 3000,NDIM,IS2                                4820
   99 NDIMH=NDIMH+NDIM                                                  4830
      SUM2H=SUM2H+SUM2                                                  4840
  100 CONTINUE                                                          4850
  103 QHZ=SUM2H/SUM2Z                                                   4860
      IS2=SUM2H+0.005                                                   4870
      IF(CHECK.GE.1) PRINT 3010, NDIMH,IS2,QHZ                          4880
      RETURN                                                            4890
 2010 FORMAT(1X,I3* . *8I3)                                             4900
 2020 FORMAT(7X,8I3)                                                    4910
 3000 FORMAT(*+*30X,I6,I16)                                             4920
 3010 FORMAT(32X,33(1H.)/31X,I6,I16,2X,E10.4)                           4930
      END                                                              4940
      SUBROUTINE SETUP(Z,M,A,V,L,ADJN)                                  4950
      INTEGER A,Z,ADJN                                                  4960
      DIMENSION A(8,8),V(8,8),L(8),ADJN(8)                              4970
      INTEGER G2(2,2),G2I(2,2),L2(2),G2A(2)                             4980
      INTEGER F4(4,4),F4I(4,4),L4(4),F4A(4)                             4990
$
$
      INTEGER E6(6,6),E6I(6,6),L6(6),E6A(6)                             5000
      INTEGER E7(7,7),E7I(7,7),L7(7),E7A(7)                             5010
      INTEGER E8(8,8),E8I(8,8),L8(8),E8A(8)                             5020
      DATA (G2(I),I=1,4)/2,-1,-3,2/                                     5030
      DATA (F4(I),I=1,16)/2,-1,0,0,-1,2,-1,0,0,-2,2,-1,0,0,-1,2/        5040
      DATA (E6(I),I=1,36)/2,-1,0,0,0,0,-1,2,-1,0,0,0,-2,2,-1,0,0,0,     5050
     1              -1,2,-1,0,0,0,0,-1,2,0,0,0,0,-1,0,0,2/             5060
      DATA (E7(I),I=1,49)/2,-1,0,0,0,0,0,-1,2,-1,0,0,0,0,-1,2,-1,0,0,0, 5070
     1              -1,0,-1,2,-1,0,0,0,0,0,-1,2,-1,0,0,0,0,0,-1,       5080
     1              2,0,0,0,-1,0,0,0,2/                                5090
      DATA (E8(I),I=1,64)/2,-1,0,0,0,0,0,0,-1,2,-1,0,0,0,0,0,-1,2,-1,   5100
     1              0,0,0,0,0,-1,2,-1,0,0,0,0,0,-1,2,-1,0,-1,          5110
     2              0,0,0,0,-1,2,-1,0,0,0,0,0,0,-1,2,0,0,0,0,0,-1,     5120
     3              0,0,2/                                             5130
      DATA (G2I(I),I=1,4)/6,3,3,2/                                      5140
      DATA (F4I(I),I=1,16)/4,6,4,2,6,12,8,4,4,8,6,3,2,4,3,2/            5150
      DATA (E6I(I),I=1,36)/4,5,6,4,2,3,5,10,12,8,4,6,6,12,18,12,6,9,4,8, 5160
     1              12,10,5,6,2,4,6,5,4,3,3,6,9,6,3,6/                 5170
      DATA (E7I(I),I=1,49)                                              5180
     X              /4,6,8,6,4,2,4,6,12,16,12,8,4,8,8,16,24,18,12,6,   5190
     1              12,6,12,18,15,10,5,9,4,8,12,10,8,4,6,2,4,6,5,      5200
     2              4,3,3,4,8,12,9,6,3,7/                              5210
      DATA (E8I(I),I=1,64)                                              5220
     X              /2,3,4,5,6,4,2,3,3,6,8,10,12,8,4,6,4,8,12,15,18,   5230
     1              12,6,9,5,10,15,20,24,16,8,12,6,12,18,24,30,20,     5240
     2              10,15,4,8,12,16,20,14,7,10,2,4,6,8,10,7,4,5,       5250
     3              3,6,9,12,15,10,5,8/                                5260
      DATA (L2(I),I=1,2)/10,6/                                          5270
      DATA (L4(I),I=1,4)/22,42,30,16/                                   5280
      DATA (L6(I),I=1,6)/16,30,42,30,16,22/                            5290
      DATA (L7(I),I=1,7)/34,66,96,75,52,27,49/                         5300
      DATA (L8(I),I=1,8)/58,114,168,220,270,182,92,136/                5310
      DATA (G2A(I),I=1,2)/1,0/,(F4A(I),I=1,4)/1,0,0,0/                 5320
      DATA (E6A(I),I=1,6)/0,0,0,0,0,1/,(E7A(I),I=1,7)/1,0,0,0,0,0,0/   5330
      DATA (E8A(I),I=1,8)/1,0,0,0,0,0,0,0/                              5340
$--THIS SUBROUTINE SETS UP THE CARTAN MATRIX A, THE SCALAR PRODUCT      5350
$--MATRIX V, THE LEVEL VECTOR L, AND THE ADJOINT REPRESENTATION FOR     5360
$--AN ALGEBRA Z OF RANK M                                              5370
C--THE ARRAYS G2,F4,E6,E7,E8                                            5380
C            - CONTAIN THE INTEGER PART OF THE CARTAN MATRIX A FOR      5390
C              THE CORRESPONDING ALGEBRA G2,F4, ETC.                    5400
C--THE ARRAYS G2I,F4I,E6I,E7I,E8I                                       5410
C            - ARE THE INVERSES , ALSO IN INTEGER FORM                  5420
C--THE ARRAYS L2,L4,L6,L7,L8                                            5430
C            - ARE THE CORRESPONDING LEVEL VECTORS                      5440
C--THE VECTORS G2A,F4A,E6A,E7A,E8A                                      5450
C            - ARE THE ADJOINT REPRESENTATIONS OF THE CORRESPONDING ALGEBRAS 5460
C                                                                       5470
C--SET UP A V L FOR ALGEBRAS E,F OR G, OR                               5480
C--SET UP PART OF L AND V FOR ALGEBRAS A,B,C OR D                       5490
      DO 70 I=1,M                                                       5500
      L(I)=I*(2*M-I)                                                    5510
      ADJN(I)=0                                                         5520
```

```
         DO 70 J=1,M                                              5530
         A(I,J)=0                                                 5540
         Y=MINO(I,J)                                              5550
         GO TO (70,70,70,70,10,50,60),Z                           5560
C--ALGEBRA TYPE E (RANK=6,7,8)                                    5570
      10 MM=M-7                                                   5580
         IF(MM) 20,30,40                                          5590
      20 A(I,J)=E6(I,J) $ L(I)=L6(I) $ ADJN(I)=E6A(I) $ Y=E6I(I,J)/3.  5600
         GO TO 70                                                 5610
      30 A(I,J)=E7(I,J) $ L(I)=L7(I) $ ADJN(I)=E7A(I) $ Y=E7I(I,J)/2.  5620
         GO TO 70                                                 5630
      40 A(I,J)=E8(I,J) $ L(I)=L8(I) $ ADJN(I)=E8A(I) $ Y=E8I(I,J)    5640
         GO TO 70                                                 5650
C--ALGEBRA TYPE F                                                 5660
      50 A(I,J)=F4(I,J) $ L(I)=L4(I) $ ADJN(I)=F4A(I) $ Y=F4I(I,J)/2.  5670
         GO TO 70                                                 5680
C--ALGEBRA TYPE G                                                 5690
      60 A(I,J)=G2(I,J) $ L(I)=L2(I) $ ADJN(I)=G2A(I) $ Y=G2I(I,J)/3.  5700
      70 V(I,J)=Y                                                 5710
         IF(Z.GT.4) RETURN                                        5720
C--SET UP THE REST OF ARRAYS A,V,L FOR THE ALGEBRAS A,B,C OR D    5730
         X=M                                                      5740
         DO 80 I=1,M                                              5750
         A(I,I)=2                                                 5760
         IF(I.LT.M) A(I,I+1)=-1                                   5770
      80 IF(I.GT.1) A(I,I-1)=-1                                   5780
         GO TO (90,110,130,150),Z                                 5790
C--ALGEBRA TYPE A                                                 5800
      90 DO 100 I=1,M                                             5810
         L(I)=(M-I+1)*I                                           5820
         DO 100 J=1,M                                             5830
         Y=MINO(M+1-I,M+1-J)*MINO(I,J)                            5840
     100 V(I,J)=Y/(X+1.)                                          5850
         ADJN(1)=ADJN(M)=1                                        5860
         IF(M.EQ.1) ADJN(1)=2                                     5870
         RETURN                                                   5880
C--ALGEBRA TYPE B                                                 5890
     110 DO 120 I=1,M                                             5900
         V(M,I)=V(M,I)*0.5                                        5910
         V(I,M)=V(I,M)*0.5                                        5920
     120 L(I)=I*(2*M-I+1)                                         5930
         L(M)=(M*(M+1))/2.                                        5940
         A(M-1,M)=-2                                              5950
         ADJN(2)=1                                                5960
         RETURN                                                   5970
C--ALGEBRA TYPE C                                                 5980
     130 A(M,M-1)=-2                                              5990
$
$
         DO 140 I=1,M                                             6000
         DO 140 J=1,M                                             6010
     140 V(I,J)=V(I,J)*0.5                                        6020
         ADJN(1)=2                                                6030
         RETURN                                                   6040
C--ALGEBRA TYPE D                                                 6050
     150 A(M-2,M)=A(M,M-2)=-1                                     6060
         A(M-1,M)=A(M,M-1)=0                                      6070
         DO 160 I=1,M                                             6080
         V(M,I)=V(M-1,I)=V(I,M)=V(I,M-1)=V(M,I)*0.5               6090
         L(I)=I*(2*M-I-1)                                         6100
     160 CONTINUE                                                 6110
         V(M-1,M-1)=V(M,M)=X/4.                                   6120
         V(M-1,M)=V(M,M-1)=(X-2.)/4.                              6130
         L(M-1)=L(M)=(M*(M-1))/2                                  6140
         ADJN(2)=1                                                6150
         RETURN                                                   6160
         END                                                      6170
         SUBROUTINE SORT1(LV,W,NW)                                6180
         INTEGER W(1),LV(1)                                       6190
C--SORT ARRAY LV(NW) IN ORDER OF INCREASING MAGNITUDE CARRYING ARRAY W  6200
$--IN PARALLEL. THE VALUES IN LV ARE ALREADY PARTIALLY SORTED.   6210
         LI=I=1                                                   6220
      20 IJ=I                                                     6230
         LJ=LI                                                    6240
         I=I+1                                                    6250
         IF(I.GT.NW) RETURN                                       6260
         LI=LV(I)                                                 6270
         IF(LJ.LE.LI) GO TO 20                                    6280
         IR=I                                                     6290
         LTM=LJ                                                   6300
```

```
   30 ITM=W(IR)                                                         6310
      W(IR)=W(IJ)                                                       6320
      W(IJ)=ITM                                                         6330
      LV(IR)=LJ                                                         6340
      IR=IJ                                                             6350
      IJ=IJ-1                                                           6360
      IF(IJ.EQ.0) GO TO 50                                             6370
      LJ=LV(IJ)                                                         6380
      IF(LJ.GT.LI) GO TO 30                                            6390
   50 LV(IR)=LI                                                         6400
      LI=LTM                                                            6410
      GO TO 20                                                          6420
      END                                                              6430
      SUBROUTINE SORT2(LV,W,MU,NW)                                     6440
      INTEGER W(1),MU(1),LV(1)                                         6450
$--SORT ARRAY LV(NW) IN ORDER OF INCREASING MAGNITUDE CARRYING ARRAYS  6460
$--W AND MU IN PARALLEL.  THE VALUES IN LV ARE ALREADY PARTIALLY SORTED. 6470
      LI=I=1                                                            6480
   20 IJ=I                                                             6490
      LJ=LI                                                            6500
      I=I+1                                                            6510
      IF(I.GT.NW) RETURN                                              6520
      LI=LV(I)                                                         6530
      IF(LJ.LE.LI) GO TO 20                                           6540
      IR=I                                                            6550
      LTM=LJ                                                          6560
   30 ITM=W(IR)                                                       6570
      W(IR)=W(IJ)                                                     6580
      W(IJ)=ITM                                                       6590
      ITM=MU(IR)                                                      6600
      MU(IR)=MU(IJ)                                                   6610
      MU(IJ)=ITM                                                      6620
      LV(IR)=LJ                                                       6630
      IR=IJ                                                           6640
      IJ=IJ-1                                                         6650
      IF(IJ.EQ.0) GO TO 50                                           6660
      LJ=LV(IJ)                                                       6670
      IF(LJ.GT.LI) GO TO 30                                          6680
   50 LV(IR)=LI                                                       6690
      LI=LTM                                                          6700
      GO TO 20                                                        6710
      END                                                            6720
      SUBROUTINE SPRD(C,P,Q,N,S)                                     6730
      INTEGER P(8),Q(8)                                              6740
      DIMENSION C(8,8),Y(8)                                          6750
$--FIND THE SCALAR PRODUCT S OF TWO VECTORS P(N),Q(N).               6760
$--C IS THE SCALAR PRODUCT MATRIX.                                   6770
      S=0                                                            6780
      DO 60 I=1,N                                                    6790
   60 Y(I)=0                                                         6800
      DO 62 I=1,N                                                    6810
      DO 62 J=1,N                                                    6820
   62 Y(I)=Y(I)+C(I,J)*P(J)                                          6830
      DO 70 I=1,N                                                    6840
   70 S=S+Y(I)*Q(I)                                                  6850
      RETURN                                                         6860
      END                                                            6870
      SUBROUTINE UNPACK(WN,M,WT)                                     6880
      INTEGER WN(8),WT,S,T                                           6890
      DATA I15/100000000000000/,I16/1000000000000000/               6900
C--THIS SUBROUTINE DOES THE REVERSE PROCESS OF SUBROUTINE PACK       6910
C--VIZ.  THE CONTENTS OF WT IS UNPACKED INTO ARRAY WN               6920
C--AND 50 IS SUBTRACTED FROM EACH COMPONENT OF WN TO OBTAIN ITS TRUE 6930
C--VALUE, EXCEPT WHEN M=1.                                           6940
      IF(M.NE.1) GO TO 120                                           6950
      WN(1)=WT                                                       6960
      RETURN                                                         6970
  120 WN(8)=0                                                        6980
      IC=0                                                           6990
$                                                                   
$                                                                   
      DO 130 I=1,10                                                  7000
      IC=IC+I16                                                      7010
      IF(IC.LE.WT) GO TO 130                                         7020
      WN(8)=10*(I-1)                                                 7030
      GO TO 140                                                      7040
  130 CONTINUE                                                       7050
C--UNPACK THE 16TH DIGIT OF WT AS 2ND DIGIT OF WN(8)                 7060
  140 T=WT-IC+I16                                                    7070
      IC=0                                                           7080
```

```
        DO 150 I=1,10                                              7090
        IC=IC+I15                                                 7100
        IF(IC.LE.T) GO TO 150                                    7110
        WN(8)=WN(8)+I-1                                          7120
        GO TO 160                                                7130
150     CONTINUE                                                  7140
C--UNPACK THE 15TH DIGIT OF WT AS THE 1ST DIGIT OF WN(8)         7150
160     T=T-IC+I15                                                7160
        IP=10                                                     7170
C--UNPACK THE REMAINING 14 DIGITS                                7180
        DO 170 IB=1,14                                           7190
        I=15-IB                                                  7200
        S=T/(10**(I-1))                                          7210
        II=(I+1)/2                                               7220
        IF(IP.EQ.1) WN(II)=WN(II)+S                             7230
        IF(IP.EQ.10) WN(I/2)=10*S                               7240
        IL=IP                                                    7250
        IF(IL.EQ.10) IP=1                                       7260
        IF(IL.EQ.1) IP=10                                       7270
170     T=T-S*(10**(I-1))                                        7280
C--SUBTRACT 50 FROM EACH COMPONENT OF WN                        7290
        DO 180 I=1,M                                             7300
180     WN(I)=WN(I)-50                                           7310
        RETURN                                                    7320
        END                                                       7330
        SUBROUTINE WTSYSZ                                         7340
        LOGICAL TEVEN                                             7350
        INTEGER ALGNAM,CHECK,Z,ZRANK,HWZ,R,A,ADJN               7360
        INTEGER W,WADJ,T                                         7370
        COMMON /COM0/ ALGNAM(7),CHECK                           7380
        COMMON /COM1/ HWZ(8),Z,ZRANK,NDIMZ,SUM2Z                7390
        COMMON /COM2/ A(8,8),V(8,8),R(8),ADJN(8)                7400
        COMMON /COM3/ WADJ(130),LVADJ(130),NWADJ,NWAMAX         7410
        COMMON /COM4/ W(5000),LV(5000),MU(5000),NW,NWMAX        7420
$--THIS SUBROUTINE CALCULATES THE RESTRICED WEIGHT SYSTEM       7430
$--(WEIGHTS W, LEVELS LV, MULTIPLICITES MU) OF THE I.R. HWZ OF Z 7440
        CALL SETUP(Z,ZRANK,A,V,R,ADJN)                          7450
$--FIND THE RESTRICTED WEIGHT SYSTEM (WADJ,LVADJ) WITH HIGHEST WEIGHT ADJN-- 7460
$--I.E. SYSTEM OF POSITIVE ROOTS.  OMIT THE ZERO WEIGHTS.       7470
        CALL GETLL(ADJN,R,ZRANK,LLADJ,T)                        7480
        LLADJ=LLADJ-1                                            7490
        CALL CALCWT(WADJ,LVADJ,ADJN,A,ZRANK,LLADJ,NWADJ,NWAMAX) 7500
        CALL SORT1(LVADJ,WADJ,NWADJ)                            7510
$--FIND THE RESTRICTED WEIGHT WYSTEM (W,LV) WITH HIGHEST WEIGHT HWZ 7520
        CALL GETLL(HWZ,R,ZRANK,LL,T)                            7530
        LLM=LL                                                   7540
        IF(MOD(T,2).NE.0) LLM=LLM+1                             7550
        CALL CALCWT(W,LV,HWZ,A,ZRANK,LLM,NW,NWMAX)             7560
        CALL SORT1(LV,W,NW)                                     7570
$--FIND THE MULTIPLICITIES OF THE WEIGHTS W AND THE DIMENSION OF HWZ 7580
        TEVEN=.FALSE.                                            7590
        IF(MOD(T,2).EQ.0) TEVEN=.TRUE.                          7600
        NDIMZ=0                                                  7610
        MUHW=1                                                   7620
        CALL FREUD(W,LV,MU,WADJ,V,NW,NWADJ,ZRANK,MUHW,NDIMZ,LL,TEVEN, 7630
       1SUM2Z)                                                   7640
        IS2=SUM2Z+0.005                                         7650
        IF(CHECK.GE.1) PRINT 1000,NDIMZ,IS2                     7660
        PRINT 1010                                               7670
        IF(CHECK.GE.1) PRINT 1020                               7680
        RETURN                                                    7690
1000    FORMAT(*+*30X,I6,I16)                                   7700
1010    FORMAT(7X,24(1H.))                                      7710
1020    FORMAT(*+*30X,34(1H.))                                  7720
        END                                                       7730
```

SYMMETRIZED KRONECKER POWERS OF
REPRESENTATIONS OF SEMISIMPLE LIE ALGEBRAS

Nigel Backhouse

1.   Introduction

The concept of a symmetrized power of a group representation arises naturally and with far-reaching implications in the
representation theory of the classical linear group.   Let the
general linear group GL(d) act via its self-representation, D say,
on the d-dimensional complex vector space V with basis $x_1, x_2, \ldots,$
x_d.   Then GL(d) acts via the n-th power of D, denoted D^n, on the
d^n-dimensional space $\otimes^n V$ of n-th rank tensors with basis

$$\{\phi_{i_1 i_2 \ldots i_n} = (x_{i_1}, x_{i_2}, \ldots, x_{i_n}) \mid i_s = 1, 2, \ldots, d \text{ for } s = 1, 2, \ldots, n\}.$$

The full permutation group S_n acts on $\otimes^n V$ by interchanging basis
element indices, and this action commutes with that of GL(d),
with the result that $\otimes^n V$ decomposes into a direct sum of GL(d)
$\times S_n$ – invariant subspaces $\otimes^{[\nu]} V$, where the label $[\nu]$ denotes
an irreducible representation of S_n. $\otimes^{[\nu]} V$ carries the representation $D^{[\nu]} \otimes [\nu]$ of GL(d) $\times S_n$; hence we may write

(1) $$D^n = \underset{[\nu]}{\otimes} f_{[\nu]} D^{[\nu]},$$

where $f_{[\nu]} = \dim [\nu]$. $D^{[\nu]}$, called the $[\nu]$ -*symmetrized power* of
D, turns out to be irreducible.   It is important to note that
$D^{[\nu]}$ vanishes if d is less than the number of rows of $[\nu]$. These
results are considered more fully in the standard applied group
theory texts, in particular those written by Hamermesh, Boerner,
Weyl, Lomont, Miller, Murnaghan [15, 5, 26, 21, 23, 24].

Now equation (1) still makes sense if D is any finite-
dimensional representation of any group G, for D embeds an image
of G as a subgroup of GL(d), d = dim D, and then (1) holds by
restriction to the subgroup. In general $D^{[\nu]}$ is not irreducible,
but it is certainly a well-defined finite-dimensional representation of G and deserves attention.

Recent work on the symmetrization of the representations
of various groups is to be found in the papers listed in the
references. We gather from these articles that the interest in
symmetrized powers is much more than purely mathematical, for
there are many important applications particularly to the quantum
mechanics of identical particles. In this paper we discuss the
symmetrization of finite-dimensional representations of semisimple
Lie algebras using weight diagram techniques. The results of
Andersen [1] and our own experience shows that in this field, hand
calculations, though elementary, soon become very laborious. The
nature of the simple counting methods involved clearly indicate
that the problem is set up for analysis by computer.

2.   The Frobenius Formula and Its Applications

If χ is the character of the representation D, then it is
shown in Weyl [26] that the character of $D^{[\nu]}$ is given by the
Frobenius formula

$$(2) \qquad \chi^{[\nu]}(g) = \sum_{(\mu)} \frac{\chi(g)^{\mu_1} \chi(g^2)^{\mu_2} \ldots \chi(g^n)^{\mu_n}}{\alpha(\mu)} [\nu](\mu)$$

for $g \in G$, where (μ) denotes the conjugacy class of S_n with
cycle structure $(1^{\mu_1} 2^{\mu_2} \ldots n^{\mu_n})$ and

$$\alpha(\mu) = 1^{\mu_1} 2^{\mu_2} \ldots n^{\mu_n} \mu_1! \mu_2! \ldots \mu_n!$$

is the order of the centralizer of an element of (μ). (Exercise: use equation (2) to establish the conjecture in [17].)

This formula was used by Boyle [6] and Andersen [1] to symmetrize representations of molecular point groups and low-dimensional semisimple Lie groups, respectively. Unfortunately, formula (2), being quite general, is not necessarily ideally suited to the most efficient calculation in any given case. In fact, the compilation of the tables in [1,6] must have involved a tremendous amount of work. In a recent paper [3] we showed how to symmetrize representations of the molecular point groups, taking advantage of the particular structure of the latter.

To go beyond Andersen [1], we begin with the group SU(2) and observe that the restriction of D(j) (its $2j + 1$-dimensional irreducible representation) to a maximal torus, the circle group U(1), is equivalent to the direct sum $\overset{j}{\underset{k=-j}{\otimes}} \psi_k$, where $\theta \rightarrow \psi_k(\theta) = e^{ik\theta}$ defines a 1-dimensional unitary representation of U(1). It is easy to see that symmetrizing D(j) is equivalent to symmetrizing the representation $\overset{j}{\underset{k=-j}{\otimes}} \psi_k$ of U(1) and then lifting back to SU(2). Now it can be shown from the Frobenius formula (2) that for representations $D_1, D_2, \ldots D_r$ we have the symmetrized multinomial theorem

$$(3) \quad \overset{r}{\underset{i=1}{\oplus}} D_i^{[\nu]} = \oplus \{\sigma([\nu]; [\nu_1], [\nu_2], \ldots, [\nu_r]) \overset{r}{\underset{i=1}{\otimes}} D_i^{[\nu_1]}\}$$

where $\sigma([\nu]; [\nu_1], [\nu_2], \ldots, [\nu_r])$ is the frequency of $[\nu]$ in $[\nu_1] \otimes [\nu_2] \otimes \cdots \otimes [\nu_r] \uparrow S_n = [\nu_1] \odot [\nu_2] \odot \cdots \odot [\nu_r]$ (outer product), or, equivalently, the frequency of $[\nu_1] \otimes [\nu_2] \otimes \cdots \otimes [\nu_r]$ in $[\nu] \downarrow S_{n_1} \times S_{n_2} \times \cdots \times S_{n_r}$, and the sum in (3) is over all irreducibles $[\nu_i]$ of S_{n_i}, $i = 1, 2, \ldots, r$, for all partitions $n_1 + n_2 + \cdots n_r = n$. This is the group theoretic version of a generalization of result III on p. 290 of Littlewood [20]. Equation (3) is particularly useful in our case since all summands vanish except those which correspond to $[\nu_i] = [n_i]$, the totally symmetric or trivial representation of S_{n_i}, for all i. So the determination of the σ's depends on knowledge of outer direct products of trivial representations of the symmetric groups, which is a very simple counting problem -- for the rules of the game see Hamermesh [15]. The uncomplicated nature of the representations of SU(2) allows not only the easy direct application of equation (3) but also the derivation of a whole host of formulae, both new and old, connecting various symmetrized powers of the D(j)'s. These are explored in some detail in [13].

The extension of these ideas to other semisimple Lie groups, and then to their Lie algebras, is that in the compact case, if D is the unitary irreducible representation to be symmetrized, then its restriction to a maximal torus is equivalent to a representation of the form $\phi \rightarrow \bigoplus_{\lambda \in \Lambda} \exp(i\omega_\lambda \cdot \phi)$, where $\phi = (\phi_1, \phi_2, \ldots, \phi_r)$, r = rank G, are canonical coordinates for the maximal torus, and the vectors ω_λ are weights labelled by λ taken from some index set Λ. Multiplicities of weights can occur so it is possible for ω_λ to equal ω_λ', with $\lambda \neq \lambda'$. Now we deduce from (3), for given $[\nu]$, that $D^{[\nu]}$ contains the weight $\sum_{\lambda \in \Lambda} n_\lambda \omega_\lambda$ with multiplicity $\sigma([\nu]; [n_\lambda], \text{all } \lambda \in \Lambda)$, the latter being the frequency of $[\nu]$ in $\underset{\lambda \in \Lambda}{\odot} [n_\lambda]$, where $[n_\lambda]$ is the trivial representation of S_{n_λ} and

$\sum_{\lambda \in \Lambda} n_\lambda = n$. Here we have applied (3) in the case that each D_i is 1-dimensional. It is also possible to choose the D_i's in the form $\phi \rightarrow m_\lambda \exp(i\omega_\lambda \cdot \phi)$, where m_λ is the multiplicity of the weight ω_λ. But we then have to contend with the problem that some of the $[\nu_i]$ are non-trivial.

The above has been derived from (3) on the assumption that G is compact. However, since the final answer is given in terms of weight vectors, it is perfectly applicable to the case that D is a representation, preferably finite-dimensional, with weights $\{\underset{\sim}{\omega}_\lambda \mid \lambda \varepsilon \Lambda\}$, of a semisimple Lie algebra.

The remainder of the paper is devoted to hand calculations for simple examples. We begin slowly to gain facility in the use of our method.

3. The Algebra A_2

In order to discuss the weight space of A_2 we introduce the fundamental dominant weights $\underset{\sim}{\alpha}_1 = 1/6 (\sqrt{3}, 1)$ and $\underset{\sim}{\alpha}_2 = 1/6 (\sqrt{3}, -1)$. The weights of finite-dimensional representations of A_2 are integral linear combinations $\mu_1 \underset{\sim}{\alpha}_1 + \mu_2 \underset{\sim}{\alpha}_2$ of the basis, and are termed *dominant* if μ_1, μ_2 are both nonnegative. Thus dominant weights lie in the interior or boundary of a so-called *Weyl chamber*, which by application of the Weyl group gives the whole of weight space. It is clear that a representation is determined by its dominant weights. Furthermore, an irreducible representation (λ_1, λ_2) can be labelled by its highest weight $\lambda_1 \underset{\sim}{\alpha}_1 + \lambda_2 \underset{\sim}{\alpha}_2$, which is a maximal distance from the origin. Finally, a notational point: $\bar{n} (\mu_1 \underset{\sim}{\alpha}_1 + \mu_2 \underset{\sim}{\alpha}_2)$ means the weight $\mu_1 \underset{\sim}{\alpha}_1 + \mu_2 \underset{\sim}{\alpha}_2$ occurs n times.

The Quark Representation (1.0)

This 3-dimensional representation, denoted $\underset{\sim}{3}$ in the physics literature, has weights $\underset{\sim}{\alpha}_1, -\underset{\sim}{\alpha}_2, \underset{\sim}{\alpha}_2 - \underset{\sim}{\alpha}_1$, which we display in Figure 1.

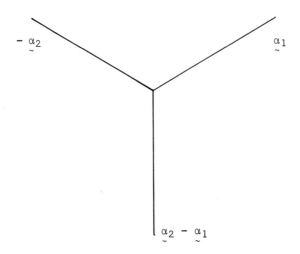

Figure 1.   The Weight Diagram of (1,0).

For n = 2 we have the partitions $2 + 0 + 0 = 2$, $1 + 1 + 0 = 2$, and permutations of these. Now $[2] \odot [0] \odot [0] = [2]$ and $[1] \odot [1] \odot [0] = [2] \oplus [1^2]$, so $(1,0)^{[2]}$ and $(1,0)^{[1^2]}$ have respective weights $\{2\underset{\sim}{\alpha}_1, -2\underset{\sim}{\alpha}_2, 2\underset{\sim}{\alpha}_2 - 2\underset{\sim}{\alpha}_1, \underset{\sim}{\alpha}_1 - \underset{\sim}{\alpha}_2, \underset{\sim}{\alpha}_2, -\underset{\sim}{\alpha}_1\}$ and $\{\underset{\sim}{\alpha}_1 - \underset{\sim}{\alpha}_2, -\underset{\sim}{\alpha}_1, \underset{\sim}{\alpha}_2\}$. To obtain these we have that, for $(1,0)^{[2]}$, any two weights can be added together, but for $(1,0)^{[1^2]}$, two weights can be added together only if different. $(1,0)^{[2]}$ has dominant weights $2\underset{\sim}{\alpha}_1$, $\underset{\sim}{\alpha}_2$, of which $2\underset{\sim}{\alpha}_1$ is the higher. Thus $(1,0)^{[2]}$ is the sextet 6 or (2,0). $(1,0)^{[1^2]}$ has dominant weight $\underset{\sim}{\alpha}_2$, so is the anti-quark representation $\underset{\sim}{\bar{3}}$ or (0,1).

For $n = 3$ we have the partitions $3 + 0 + 0 = 3$, $2 + 1 + 0 = 3$
and $1 + 1 + 1 = 3$. Now $[3] \odot [0] \odot [0] = [3]$, $[2] \odot [1] \odot [0] =$
$[3] \oplus [2,1]$ and $[1] \odot [1] \odot [1] = [3] \oplus \bar{2} [2,1] \oplus [1^3]$.   We
therefore find that $(1,0)^{[3]}$, $(1,0)^{[2,1]}$ and $(1,0)^{[1^3]}$ have
respective weights $\{3\alpha_1, - 3\alpha_2, 3\alpha_2 - 3\alpha_1, 2\alpha_1 -\alpha_2, \alpha_1 - 2\alpha_2,$
$- \alpha_2 - \alpha_1, \alpha_2 - 2\alpha_1, 2\alpha_2 - \alpha_1, \alpha_2 + \alpha_1, 0\}, \{2\alpha_1 - \alpha_2, \alpha_1 - 2\alpha_2,$
$- \alpha_2 - \alpha_1, \alpha_2 - 2\alpha_1, 2\alpha_2 - \alpha_1, \alpha_2 + \alpha_1, \bar{2} 0\}, \{0\}$.   These have
respective dominant weights $\{3\alpha_1, \alpha_1 + \alpha_2, 0,\}$, $\{\alpha_1 + \alpha_2, \bar{2} 0\}$,
$\{0\}$.   Thus $(1,0)^{[3]} = (3,0)$ (decuplet 1 0); $(1,0)^{[2,1]} = (1,1)$
(the famous octet 8); $(1,0)^{[1^3]} = (0,0)$ (the trivial or zero
representation).

For $n = 4$ we find explicitly $(1,0)^{[4]} = 15'$ or $(4,0)$;
$(1,0)^{[3,1]} = \underline{15}$ or $(2,1)$; $(1,0)^{[2,1^2]} = \bar{3}$ or $(0,1)$; $(1,0)^{[2,2]}$
$= \bar{6}$ or $(0,2)$. More generally, $(1,0)^{[\mu_1, \mu_2, \mu_3]}$ has highest
weight $\mu_1\alpha_1 + \mu_2(\alpha_2 - \alpha_1) + \mu_3 (-\alpha_2) = (\mu_1 - \mu_2)\alpha_1 + (\mu_2 -\mu_3)\alpha_2$,
so it contains $(\mu_1 - \mu_2, \mu_2 - \mu_3)$. In fact it is well-known from
the constructive theory of representations of A_2 that
$(1,0)^{[\mu_1,\mu_2,\mu_3]} = (1,0)^{[\mu_1 - \mu_3, \mu_2 - \mu_3]} = (\mu_1 - \mu_2, \mu_2 - \mu_3)$.
As an aside we note that this can be used to link representations
of $SU(3)$ (compact real form of A_2) with those of $SO(3)$. For
consider $(\lambda_1, \lambda_2) \downarrow (SO(3): (\lambda_1,\lambda_2) = (1,0)^{[\lambda_1+\lambda_2,\lambda_2]}$ and
$(1,0) \downarrow SO(3) = D(1)$. Thus $(\lambda_1,\lambda_2) \downarrow SO(3) = D(1)^{[\lambda_1+\lambda_2,\lambda_2]}$. The

latter has been derived in [13], and we have

(4) $D^{[\lambda_1 + \lambda_2]} \oplus D^{[\lambda_1 + \lambda_2 - 1]} \oplus \ldots \oplus D^{[\lambda_1]} - \{D^{[\lambda_2 - 1]} \oplus \ldots \oplus D^{[1]} \oplus D^{[0]}\},$

where $D = D(1)$ and $D^{[\mu]} = D(\mu) \oplus D(\mu - 2) \oplus \ldots \oplus D(1)$ or $D(0)$.

This agrees with the conclusion of Theorems 5.1 and 5.2 of [9].

The Sextet Representation (2,0)

The weights of (2,0) are $\{2\alpha_1, -2\alpha_2, \ 2\alpha_2 - 2\alpha_1, \ \alpha_1 - \alpha_2,$
$\alpha_2, \ -\alpha_1\}$ displayed in Figure 2.

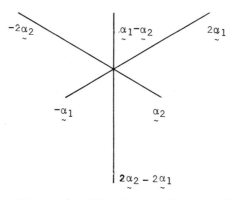

Figure 2.   The Weight Diagram of (2,0).

The weights of $(2,0)^{[2]}$ are sums of pairs of weights of (2,0).
Thus the dominant weights of $(2,0)^{[2]}$ are $\{4\alpha_1, \ 2\alpha_2, \ 2\alpha_2, \ 2\alpha_1 +$
$\alpha_2, \ \alpha_1, \ \alpha_1\}$. This does not give an irreducible representation,
but is the sum of (4,0) and the other sextet (0,2) with dominant
weights $\{4\alpha_1, \ 2\alpha_2, \ 2\alpha_1 + \alpha_2, \ \alpha_1\}$ and $\{2\alpha_2, \ \alpha_1\}$, respectively.
The dominant weights of $(2,0)^{[1^2]}$ are $\{2\alpha_2, \ 2\alpha_1 + \alpha_2, \ \alpha_1, \alpha_1\}$,
which gives (2,1).

The weights of $(2,0)^{[1^3]}$ are sums of triplets of distinct
weights of (2,0). The dominant weights are $\{0, \ 3\alpha_2, \ 3\alpha_1, \ \alpha_1 +$
$\alpha_2, \ \alpha_1 + \alpha_2, \ 0\}$ which gives (3,0) $\oplus$ (0,3).

The Octet Representation (1,1)

The weights of $(1,1)$ are $\{\alpha_1 + \alpha_2,\ -\alpha_1 - \alpha_2,\ 2\alpha_2 - \alpha_1,\ \alpha_1 - 2\alpha_2,\ 2\alpha_1 - \alpha_2,\ \alpha_2 - 2\alpha_1,\ 0,\ 0\}$, displayed in Figure 3.

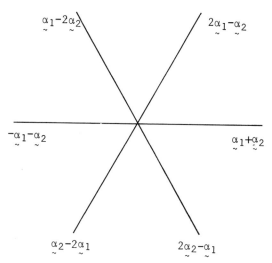

Figure 3.   The Weight Diagram of $(1,1)$.
The zero weight has multiplicity two.

The dominant weights of $(1,1)^{[2]}$ are $\{2\alpha_1 + 2\alpha_2,\ 0,\ 0,\ 0,\ 3\alpha_2,\ 3\alpha_1,\ \alpha_1 + \alpha_2,\ \alpha_1 + \alpha_2,\ 0,\ \alpha_1 + \alpha_2,\ 0,\ 0\}$, which gives $(2,2) \oplus (1,1) \oplus (0,0)$. The dominant weights of $(1,1)^{[1^2]}$ are the same, but omitting $2\alpha_1 + 2\alpha_2$ and two copies of 0. Thus we get $(3,0) \oplus (0,3) \oplus (1,1)$. The dominant weights of $(1,1)^{[1^3]}$ are $\{0,\ 0,\ \alpha_1 + \alpha_2,\ 2\alpha_1 + 2\alpha_2,\ 3\alpha_2,\ 3\alpha_2,\ \alpha_2 + \alpha_1,\ 3\alpha_1,\ 3\alpha_1,\ \alpha_1 + \alpha_2,\ 0,\ 0,\ 0,\ \alpha_1 + \alpha_2,\ \alpha_1 + \alpha_2,\ 0,\ 0\}$ which gives $(2,2) \oplus (3,0) \oplus (0,3) \oplus (1,1) \oplus (0,0)$.

4.   The Algebra B_2 (C_2)

The algebras B_2 and C_2 are isomorphic, being a reflection of the local isomorphism between the compact Lie groups 0_5 and Sp_2. B_2 has fundamental dominant weights $\underset{\sim}{\alpha}_1 = \frac{1}{2\sqrt{3}} (1,1)$ and $\underset{\sim}{\alpha}_2 = \frac{1}{2\sqrt{3}} (1,0)$. The weights of the corresponding fundamental representations are displayed in Figure 4.

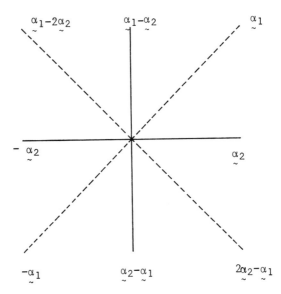

Figure 4   The Weight Diagrams for $(1,0)$

(dotted lines) and $(0,1)$.

$(1,0)$ has a zero weight.

Thus $(1,0)$ has weights $\{\underset{\sim}{\alpha}_1, -\underset{\sim}{\alpha}_1, 2\underset{\sim}{\alpha}_2 - \underset{\sim}{\alpha}_1, \underset{\sim}{\alpha}_1 - 2\underset{\sim}{\alpha}_2, 0\}$, whereas $(0,1)$ has weights $\{\underset{\sim}{\alpha}_2, -\underset{\sim}{\alpha}_2, \underset{\sim}{\alpha}_1 - \underset{\sim}{\alpha}_2, \underset{\sim}{\alpha}_2 - \underset{\sim}{\alpha}_1\}$.

$(1,0)^{[2]}$ has dominant weights $\{2\underset{\sim}{\alpha}_1, 0, 0, 2\underset{\sim}{\alpha}_2, \underset{\sim}{\alpha}_1, 0\}$, which gives $(2,0) \oplus (0,0)$. $(1,0)^{[1^2]}$ has dominant weights $\{0, 2\underset{\sim}{\alpha}_2, \underset{\sim}{\alpha}_1, 0\}$ which gives $(0,2)$. We similarly find $(1,0)^{[3]} = (3,0) \oplus (1,0)$ and $(1,0)^{[1^3]} = (0,2)$. In fact it can easily be shown by induction that $(1,0)^{[n]} = (n,0) \oplus (n-2, 0) \oplus \cdots \oplus$ $(1,0)$ or $(0,0)$.

We also find $(0,1)^{[2]} = (0, 2)$; $(0, 1)^{[1^2]} = (1,0) \oplus (0,0)$; $(0,1)^{[3]} = (0,3)$; $(0,1)^{[1^3]} = (0,1)$; $(0,1)^{[2,1]} = (1,1) \oplus (0,1)$. Generally $(0,1)^{[n]} = (0,n)$.

5.   The Algebra $A_1 \times A_1$

This algebra is that of the Lie group $SO(4)$. We denote the representations of $A_1 \times A_1$ by (n_1, n_2), which can be written as $(n_1,0) \otimes (0,n_2)$, where $(n,0)$ and $(0,n)$ both restrict to the $n + 1$-dimensional representation of A_1. Using weight vector arguments it is easy to find the following decomposition formula for symmetrizing the 4-dimensional spinor representation $(1,1)$:

(5) $(1,1)^{[\nu]} = \bigoplus_{m \geq m'} \bar\sigma([\nu]; \; [\mu],[\mu'])\{m,m'\} \otimes (0,\mu_1 - \mu_2) \otimes (0,\mu_1' - \mu_2')$

where $\sigma([\nu]; [\mu],[\mu'])$ is the frequency of $[\nu]$ in $[\mu] \odot [\mu'] = [\mu]$ $\otimes [\mu'] \uparrow S_n$, induced from the group $S_m \times S_m'$, $m + m' = n$, and where $[\mu] = [\mu_1, \mu_2]$, $[\mu'] = [\mu_1', \mu_2']$ are two-rowed Young tableau. Also, $\{m,m'\}$ is $(m - m',0) - (m - m' - 2,0)$ if $m \geq m' + 2$, is $(1,0)$ if $m' = m + 1$ and is $(0,0)$ if $m = m'$.

Using equation (5) we find the following

(6) $(1,1)^{[n]} = (n,n) \oplus (n - 2, n - 2) \oplus \cdots \oplus (1,1)$ or $(0,0)$;

(7) $(1,1)^{[n-1,1]} = \{(n, n-2) \oplus (n-2, n-2) \oplus (n-2,n)\}$
$\oplus\{(n-2,n-4) \oplus (n-4,n-4) \oplus (n-4,n-2)\} \oplus \cdots$
$\oplus\{(3,1) \oplus (1,1) \oplus (1,3)\}$ or $\{(2,0) \oplus (0,2)\}$;

(8) $(1,1)^{[n-2,1^2]} = \{(n-2, n-4) \oplus (n-2, n-2) \oplus (n-4, n-2)\}$
$\oplus\{(n-4, n-6) \oplus (n-4, n-4) \oplus (n-6, n-4)\}$
$\oplus \cdots \oplus (1,1)$ or $\{(2,0) \oplus (2,2) \oplus (0,2)\}$;

(9) $(1,1)^{[n-3,1^3]} = (n-4,\ n-4) \oplus (n-6,\ n-6) \oplus \cdots \oplus (1,1)$ or $(0,0)$

This last result is predictable using (6) since we have the result $(1,1)^{[\lambda_1,\ \lambda_2,\ \lambda_3,\ \lambda_4]} = (1,1)^{[\lambda_1-\lambda_4,\lambda_2-\lambda_4,\lambda_3-\lambda_4]}$. The above are additional to those formulae found by Gard [12].

Two other methods of symmetrization are worth mentioning. $(1,1)$ can be lifted to A_3, symmetrized, and then reduced using a branching rule. To deal with (n_1,n_2) we only have to recall that it can be written as $(n_1,0) \otimes (0,n_2)$ and then employ Littlewood's formula for the symmetrization of direct products.

(10) $(D_1 \otimes D_2)^{[\nu]} = \oplus \, \rho([\nu]; [\nu_1],[\nu_2]) \; D_1^{[\nu_1]} \otimes D_2^{[\nu_2]}$

where $[\nu]$, $[\nu_1]$, $[\nu_2]$ are representations of S_n and ρ is the frequency of $[\nu]$ in the inner Kronecker product $[\nu_1] \otimes [\nu_2]$. In the application of the formula (10) we have that D_1, D_2 are essentially representations of A_1. We also remark that (10) can be deduced from (2).

6.   Conclusion

We have shown using simple examples how representations of semisimple Lie algebras can be symmetrized in a straightforward manner. An obvious virtue of our method is that it can be easily stated in terms of the operations of counting, to obtain the σ's, and of addition and subtraction, to obtain the dominant weights. Thus the method is suitable for computer analysis and can be used in conjunction with existing programs for calculating inner multiplicities [18]. In particular it fits in with the root-diagram and weight-vector view of semisimple Lie algebras, in contrast to the view expressed through the plethysm calculus of Littlewood. Furthermore, it can be used to tackle the exceptional algebras.

We finally note that, in contrast to a direct application of formula (2), there is no need for knowledge of symmetric group representations. In this sense our method is self-contained.

Acknowledgement

The author wishes to acknowledge many valuable conversations with Patricia Gard.

References

[1]   C.M. Andersen, *Clebsch-Gordan series for symmetrized tensor products*, J. Math. Phys. 8(1967), 988-997.

[2]   N. B. Backhouse, *A comment on the reality classification of space group representations*, J. Phys. (A) 6(1973), 1115-1118.

[3]   N. B. Backhouse and P. Gard, *Symmetrized powers of point group representations*, J. Phys. (A) 7(1974), 1239-1250.

[4]   N. B. Backhouse and P. Gard, *The representation theory of the icosahedral group*, J. Phys. (A) 7(1974), 2101-2108.

[5]   H. Boerner, *Representations of Groups*, North Holland, Amsterdam, 1970.

[6]   L. L. Boyle, *The symmetrized powers of group representations*, Int. J. Quant. Chem. 6(1972), 725-746.

[7]   C. J. Bradley and B. L. Davies, *Kronecker products and symmetrized squares of irreducible representations of space groups*, J. Math.Phys. 11(1970), 1536-1552.

[8]   P. H. Butler and R. C. King, *Symmetrized Kronecker products of group representations*, Can. J. Math. 26(1974), 328-339.

[9]   A. De Baenst-Vandenbroucke, P. De Baenst and D. Speiser, *Induction procedure for the reduction $SU_n \rightarrow SO_n$*, Proc. R. Ir. Acad. 73A(1973), 131-150.

[10] P. Gard, *Symmetrized n-th powers of induced representations*, J. Phys. (A) 6(1973), 1807-1828.

[11] P. Gard, *Symmetrized n-th powers of space group representations*, J. Phys. (A) 6(1973), 1829-1836.

[12] P. Gard, *A method for obtaining symmetrized representations of SU(2) × SU(2) and the rotation group SO(4)*, J. Phys. (A) 7(1974), 2095-2100.

[13] P. Gard and N. B. Backhouse, *Methods for obtaining symmetrized representations of SU(2) and the rotation group*, J. Phys. (A) 7(1974), 1973-1803.

[14] P. Gard and N. B. Backhouse, *The reduction of symmetrized powers of corepresentations of magnetic groups*, J. Phys. (A) 8(1975), 450-458.

[15] M. Hamermesh, *Group Theory*, Addison-Wesley, Reading, Mass., 1964.

[16] K. C. Hannabuss, *Symmetrized tensor products of induced representations*, J. Phys.(A) 9(1976), 325-334.

[17] R. Higgins and D. Ballew, *An equation for finite groups*, Am. Math. Monthly 78(1971), 274-275.

[18] B. Kolman and R. E. Beck, *Computers in Lie algebras. I. Calculation of inner multiplicities*, SIAM J. Appl. Math. 25(1973), 300-312.

[19] D. Lewis, *n-th symmetrized powers of space group representations: subgroup formulae*, J. Phys. (A) 6(1973), 125-149.

[20] D. E. Littlewood, *The Theory of Group Characters and Matrix Representations of Groups*, Clarendon Press, Oxford, 1950.

[21] J. S. Lomont, *Applications of Finite Groups*, Academic Press, New York, 1970.

[22] G. W. Mackey, *Symmetric and anti-symmetric Kronecker squares and intertwining numbers of induced representations of finite groups*, Am. J. Math. 75(1953), 387-405.

[23] W. Miller, *Symmetry Groups and Their Applications*, Academic Press, New York, 1973.

[24] F. D. Murnaghan, *The Unitary and Rotation Groups*, Spartan Books, Washington, D. C., 1962.

[25] F. D. Murnaghan, *Powers of representations of the rotation group: their symmetric, alternating and other parts*, Proc. Nat. Acad. Sci. (U.S.A.) 69(1972), 1181–1184.

[26] H. Weyl, *The Theory of Groups and Quantum Mechanics*, Dover, New York, 1950.

Department of Applied Mathematics and Theoretical Physics
University of Liverpool
P. O. Box 147
Liverpool, L69   3BX

Index